AF785201

Earth Observation, Public Health and One Health Activities, Challenges and Opportunities

FSC
www.fsc.org
MIX
Paper from responsible sources
FSC® C022174

Earth Observation, Public Health and One Health

Activities, Challenges and Opportunities

Prepared by:

Public Health Agency of Canada and Canadian Space Agency

Editors:

Stéphanie Brazeau and Nicholas H. Ogden

Canadian Space Agency | Agence spatiale canadienne

Public Health Agency of Canada | Agence de la santé publique du Canada

CABI

CABI is a trading name of CAB International

CABI
Nosworthy Way
Wallingford
Oxfordshire OX10 8DE
UK

Tel: +44 (0)1491 832111
Fax: +44 (0)1491 833508
E-mail: info@cabi.org
Website: www.cabi.org

CABI
WeWork
One Lincoln St
24th Floor
Boston, MA 02111
USA

Tel: +1 (617)682-9015
E-mail: cabi-nao@cabi.org

A catalogue record for this book is available from the British Library, London, UK.

Library of Congress Cataloging-in-Publication Data

Names: Public Health Agency of Canada, author. | Brazeau, Stéphanie, editor. | Ogden, Nicolas, editor. | Canadian Space Agency, author.
Title: Earth observation, public health and one health : activities, challenges and opportunities / prepared by: Public Health Agency of Canada and Canadian Space Agency ; editors: Stéphanie Brazeau and Nicolas Ogden.
Description: Oxfordshire ; Boston, MA : CABI, [2022] | Includes bibliographical references and index. | Summary: "The potential for Earth Observation (EO) to contribute to public health practice: to understand, track, predict and manage infectious diseases and challenges of using EO data for public health purposes. Major foci are mosquito-borne diseases, tick-borne diseases, air quality and heat, water-borne diseases, vulnerable populations and pandemics (including COVID-19)"-- Provided by publisher.
Identifiers: LCCN 2021054123 (print) | LCCN 2021054124 (ebook) | ISBN 9781800621183 (hardback) | ISBN 9781800621190 (ebook) | ISBN 9781800621206 (epub)
Subjects: LCSH: One Health (Initiative) | Public health--Geographic information systems--Canada. | Public health--Research--Canada. | Public health--Canada--Methodology. | Environmental monitoring--Remote sensing.
Classification: LCC RA566.5.C2 P83 2022 (print) | LCC RA566.5.C2 (ebook) | DDC 362.10971--dc23/eng/20220121
LC record available at https://lccn.loc.gov/2021054123
LC ebook record available at https://lccn.loc.gov/2021054124

References to Internet websites (URLs) were accurate at the time of writing.

ISBN-13: 9781800621183 (hardback)
9781800621190 (OA ePDF)
9781800621206 (OA ePub)

DOI: 10.1079/9781800621183.0000

Commissioning Editor: David Hemming
Editorial Assistant: Lauren Davies
Production Editor: James Bishop

Typeset by SPi, Pondicherry, India
Printed and bound in the UK by Severn, Gloucester

Contents

List of Tables

List of Figures

List of Contributors

Stéphanie Brazeau Head of the Public Health Geomatics Unit, Public Health Risk Sciences Division, National Microbiology Laboratory, Public Health Agency of Canada, Saint-Hyacinthe, Québec, Canada
Corresponding author at: Public Health Risk Sciences Division, National Microbiology Laboratory, Public Health Agency of Canada, 3200 Sicotte, Saint-Hyacinthe, QC J2S 2M2, Canada
E-mail: stephanie.brazeau@phac-aspc.gc.ca.ca

Nicholas H. Ogden Director, Public Health Risk Sciences Division, National Microbiology Laboratory, Public Health Agency of Canada, Saint-Hyacinthe, Québec, Canada

Guy Aubé Earth Observation Applications, Space Utilization, Canadian Space Agency, Saint-Hubert, Québec, Canada

Philippe Berthiaume Public Health Risk Sciences Division, National Microbiology Laboratory, Public Health Agency of Canada, Saint-Hyacinthe, Québec, Canada

André Beaudoin Canadian Forest Service, Natural Resources Canada, Québec City, Québec,, Canada

Dominique J. Bicout Biomathematics & Epidemiology, EPSP, TIMC-IMAG, CNRS Grenoble Alpes University, VetAgro Sup, Veterinary campus of Lyon, Marcy l'Etoile, France

Marion Borderon Department of Geography and Regional Research, University of Vienna, Vienna, Austria

Catherine Bouchard Public Health Risk Sciences Division, National Microbiology Laboratory, Public Health Agency of Canada, Saint-Hyacinthe, Québec, Canada

Thibault Catry Espace Dev, Institut de Recherche pour le Développement (IRD), Montpellier, France

Marie-Josée Champagne Public Health Risk Sciences Division, National Microbiology Laboratory, Public Health Agency of Canada, Saint-Hyacinthe, Québec, Canada

Jack Chen Air Quality Research Division, Environment and Climate Change Canada, Ottawa, Ontario, Canada

Andrew Davidson Science and Technology Branch, Agriculture and Agrifood Canada, Ottawa, Canada

Didier Davignon Canadian Meteorological Centre Operations, Environment and Climate Change Canada, Dorval, Québec, Canada

Richard A. Fournier Department of Applied Geomatics, Université de Sherbrooke, Québec, Canada

Michael Gill NatureServe, Arlington, Virginia, USA

John Haynes Health and Air Quality Applications, Applied Sciences Program, NASA Earth Science Division, Washington, DC, USA

Serge Olivier Kotchi Public Health Risk Sciences Division, National Microbiology Laboratory, Public Health Agency of Canada, Saint-Hyacinthe, Québec, Canada
Ramesha S. Krishnamurthy Health Systems and Innovation Cluster, World Health Organization, Geneva, Switzerland
Antoinette Ludwig Public Health Risk Sciences Division, National Microbiology Laboratory, Public Health Agency of Canada, Saint-Hyacinthe, Québec, Canada
Jaime Martinez-Urtaza Department of Genetics and Microbiology, Universitat Autònoma de Barcelona, Barcelona, Spain
Chris McLinden Air Quality Research Division, Environment and Climate Change Canada, Toronto, Ontario, Canada
Steven Ramage Group on Earth Observation Secretariat, Geneva, Switzerland
Yves Rochon Air Quality Research Division, Environment and Climate Change Canada, Toronto, Ontario, Canada
Joacim Rocklöv Department of Public Health and Clinical Medicine, Umeå University, Umeå, Sweden
Jan C. Semenza Scientific Assessment, SRS, European Centre for Disease Prevention and Control (ECDC), Solna, Sweden
Christopher Sioris Air Quality Research Division, Environment and Climate Change Canada, Toronto, Ontario, Canada
Joaquin Trinanes University of Santiago de Compostela, CRETUS Institute, Santiago de Compostela, Spain/University of Miami, Cooperative Institute for Marine and Atmospheric Studies, Miami, Florida, USA/NOAA, Atlantic Oceanographic and Meteorological Laboratory, Miami, Florida, USA.
Juli Trtanj One Health Lead, Climate and Weather Extremes Integration Lead, NOAA Climate Program Office, Washington, DC, USA
Cécile Vignolles Centre National d'Etudes Spatiales, Directorate for Innovation, Applications and Science, Earth Observation Programme, Toulouse, France
Dirk Werle Ærde Environmental Research, Halifax, Nova Scotia, Canada
Joanne C. White Canadian Forest Service, Natural Resources Canada, Victoria, British Columbia, Canada
Michael A. Wulder Canadian Forest Service, Natural Resources Canada, Victoria, British Columbia, Canada

List of Acronyms and Abbreviations

ACCENT	Atmospheric Composition Change, the European Network of Excellence
ACI	annual crop inventory (Agriculture and Agri-food Canada)
ACIM	annual crop inventory maps
AHI	Advanced Himawari Imager
ALOS	Advanced Land Observing Satellite
AltiKa	High resolution altimeter including bi-frequency radiometric function
AMSU	Advanced Microwave Sounding Unit
ANR	Agence Nationale de la Recherche/National Research Agency (France)
ANSD	Agence Nationale de la Statistique et de la Démographie/ National Agency of Statistics and Demography (Dakar, Senegal)
AOD	aerosol optical depth
AQI	air quality index
AQUA	Aqua Earth-observing satellite mission
AQT	Association Québecoise de Télédétection/Quebec Association of Remote Sensing
ARD	analysis-ready data
ARSET	Applied remote sensing training program
ASAR	Advanced Synthetic Aperture Radar
ASDP	AirNow Satellite Data Processor
AVHRR	advanced very high-resolution radiometer
AWiF	Advanced Wide Field Sensor
AWS	Amazon Web Services
BIOMASS	biomass monitoring mission for carbon assessment
C2C	Composite to Change
CBD	Convention on Biological Diversity
CD	census district
CDC	Centers for Disease Control
CEOS	Committee on Earth Observation Satellites
CFS	Canadian Forest Service
CHRS	Center for Hydrometeorology and Remote Sensing (University of California)
CME	Climatological, Meteorological and Environmental
CMORPH	Climate Prediction Center morphing method

CNES	Centre National d'Études Spatiales/National Centre for Space Studies (France)
CoP	Community of Practice
COVID-19	coronavirus disease 2019
CPC	Climate Prediction Center
CREST	Committee for Scientific and Technical Research (EU)
CSA	Canadian Space Agency
CSE	Centre de Suivi Ecologique/Ecological Monitoring Centre
DEM	Digital Elevation Model
DFO	Department of Fisheries and Oceans (Canada)
DMSP	Defence Meteorological Satellite Program
E3	European Environment and Epidemiology
ECDC	European Centre for Disease Control
ECMWF	European Centre for Medium-Range Weather Forecasts
ECVs	Essential Climate Variables
EHA	EcoHealth Alliance
EO	Earth Observation
EOS	Earth Observation Satellites
EO4SDG	Earth Observation for Sustainable Development Goals
EPA	Environmental Protection Agency
ESA	European Space Agency
ETM+	Enhanced Thematic Mapper Plus
EVI	Enhanced Vegetation Index
FAO	Food and Agriculture Organization of the United Nations
FEER	Fire Energetics and Emissions Research
FFI	forest fragmentation index
FINN	Fire INventory from NCAR
FT	forest type
GCOM-W-AMSR2	Global Change Observation Mission – Water "Shizuku" – Advanced Microwave Scanning Radiometer 2
GDP	gross domestic product
GEO	Group on Earth Observations
GEO BON	Group on Earth Observations Biodiversity Observation Network
GEOSS	Global Earth Observation System of Systems
GFAS	Global Fire Assimilation System
GFES	Global Fire Emissions Database
GHHIN	Global Heat Health Information Network
GHRF	Global Health Risk Framework
GICC	Global Inventory for Chemistry-Climate
GIS	geographic information system
GloCover	Global Land Cover program from MERIS
GPM	Global Precipitation Measurement mission
GOCART	Goddard Chemistry Aerosol Radiation and Transport
GOES	Geostationary Satellite Server
GPSDD	Global Partnership for Sustainable Development Data
GSMaP	Global Satellite Mapping of Precipitation
HAQAST	Health and Air Quality Applied Sciences Team
HBR	human biting rate
HiRi	High Resolution Optical Imager
HR	high resolution
HRG	High Resolution Geometric
IARR	internal average relative reflectance

INPE	Instituto Nacional de Pesquisas Espaciais (Brazil's National Institute of Space)
IPCC	Intergovernmental Panel on Climate Change
IRAP	International Research and Applications Project
IRBA	Infectiologie de Terrain de l'Institut de Recherche Biomédicale des Armées/Armed Forces Biomedical Research Institute (Marseille, France)
IRD	Institut de Recherche pour le Développement/Research Institute for Development (France)
JAXA	Japan Aerospace Exploration Agency
JPSS	Joint Polar Satellite System
LiDAR	light detection and ranging
LMICs	lower middle-income countries
LST	land surface temperature
LU	land use
LULC	land use and land cover
MACC	Monitoring Atmospheric Composition and Climate with the Global Fire Assimilation System
MAIA	Multi-Angle Imager for Aerosols
MBDs	mosquito-borne diseases
MEDDE	French Ministry of Ecology
MERIS	Medium Resolution Spectrometer
Metop	meteorological operational satellite
MODIS	Moderate Resolution Imaging Spectroradiometer
MSI	Multispectral instrument
NAOMI	New AstroSat Optical Modular Instrument
NASA	National Aeronautics and Space Administration
NCAR	National Center for Atmospheric Research
NDPI	Normalized Difference Pond Index
NDVI	Normalized Difference Vegetation Index
NFI	National Forest Inventory (Natural Resources Canada)
NGO	non-governmental organizations
NIHHIS	National Integrated Heat Health Information System
NLHI	normalized landscape-based hazard index
NML	National Microbiology Laboratory
NOAA	National Oceanic and Atmospheric Administration
NPP	National Polar-orbiting Partnership
OECD	Organisation for Economic Co-operation and Development
OIE	Organisation mondiale de la santé animale/World Organisation for Animal Health
OLCI	Ocean and Land Colour Instrument
OLI	Operational Land Imager
OMI	ozone monitoring instrument
OSTIA	Operational Sea Surface Temperature and Sea Ice Analysis
PACE	Plankton, Aerosol, Cloud, ocean Ecosystem
PALSAR	Phased Array L-band Synthetic Aperture Radar
PERSIANN	Precipitation Estimation from Remotely Sensed Information using Artificial Neural Networks
PHAC	Public Health Agency of Canada
PHRS	Public Health Risk Science division
PM	particulate matter
PNLP	Programme national de lutte contre le paludisme/National Programme for the Fight Against Malaria

RCM	RADARSAT Constellation Mission
RFE	African rainfall estimate
RS	remote sensing
RVF	Rift Valley fever
SAR	synthetic aperture radar
SARAL	Satellite with Argos and Altika
SARS-CoV-2	severe acute respiratory syndrome coronavirus 2
SBAs	Societal Benefit Areas
SDG	Sustainable Development Goal
SDI	spatial data infrastructure
SeaWiFS	Sea-viewing Wide Field-of-view Sensor
SEVIRI	Spinning Enhanced Visible and Infrared Imager
SLSTR	Sea and Land Surface Temperature Radiometer
SMAP	Soil Moisture Active Passive
SMOS	Soil Moisture Ocean Salinity
SoVI	social vulnerability index
SPOT	Système Pour l'Observation de la Terre/Earth Observation System
SR	spatial resolution
SRTM	Shuttle Radar Topography Mission
SSMI	Special sensor microwave imager
SST	sea surface temperature
SWIR	short-wave infrared
SWOT	Surface Water Ocean Topography
TandDEM-X	TerraSAR-X add-on for Digital Elevation Measurement
TEMPO	Tropospheric Emissions: Monitoring of Pollution
TIRS	Thermal Infrared Sensor
TM	Thematic Mapper
TMPA	TRMM Multi-satellite Precipitation Analysis
TR	temporal resolution
TRMM	Tropical Rainfall Measuring Mission
TROPOMI	Tropospheric Monitoring Instrument
UMD	University of Maryland
UN	United Nations
UN-COPUOS	United Nations Committee on the Peaceful Uses of Outer Space
UNEP	United Nations Environment Programme
UN-GGIM	United Nations Committee of Experts on Global Geospatial Information Management
UNICEF	United Nations Children's Fund
URMITE	Unité de Recherche sur les Maladies Infectieuses et Tropicales Emergentes/Research Unit on Emerging Infectious and Tropical Diseases
USGS	United States Geological Survey
VHR	very high resolution
VIIRS	Visible Infrared Imaging Radiometer Suite
VLCE	Virtual Land Cover Engine
VOCs	volatile organic compounds
WACCM	Whole Atmosphere Community Climate Model
WGCapD	Working Group on Capacity Building & Data Democracy
WHO	World Health Organization
ZPOM	zones potentially occupied by mosquitoes

Foreword

Public Health Agency of Canada (PHAC)

As the Vice President of the National Microbiology Laboratory (NML), PHAC, I am grateful for the engaged participation of national and international experts in the preparation of this book. I would particularly like to recognize our CSA colleagues for their past and future leadership and collaboration in this field and for their role as co-lead in the development of this book.

PHAC is mandated to "promote and protect Canadians' health by preventing and controlling chronic and infectious diseases and injuries as well as preparing for and responding to public health emergencies." To accomplish this, our scientists engage in collaborative research and public health studies and use innovative methods and technologies to drive early detection and rapid response to public health threats. This includes leading studies on emerging and high-consequence infectious diseases such as Ebola, Zika, and novel respiratory pathogens like COVID-19.

NML makes continuous innovation and scientific advancement a priority to better support public health response. Initiatives range from developing new laboratory diagnostics and medical countermeasures to implementing high-performance compute clusters and bioinformatic tools to analyze genomic and other -omics data. The innovative use of space-based Earth Observation (EO) data to inform model-based risk assessments presents an exciting opportunity to further augment existing capabilities.

This book represents a significant point of departure for the future development of EO to support public health applications. Implementing book recommendations will require a collaborative One Health approach, and NML looks forward to building upon the multidisciplinary partnerships developed through this process.

My sincere thanks to all who participated in the development of this book.

Dr. P. Guillaume Poliquin, MD, PhD, FRCPC
Vice President
National Microbiology Laboratory
Public Health Agency of Canada

Foreword

Canadian Space Agency

On behalf of the Canadian Space Agency (CSA), I'm pleased to present this book on the contributions of Earth Observation (EO) to public health practices and on the challenges and opportunities of this multi-sector endeavor. I would like to recognize the leadership of the experts from the National Microbiology Laboratory of the Public Health Agency of Canada (PHAC) in their role as co-lead in the development and coordination of this manuscript on the emergence of EO and public health applications.

The mandate of the CSA is to promote the peaceful use and development of space, advance the knowledge of space through science, and ensure that space science and technology provide social and economic benefits. The CSA is delivering on this mandate in collaboration with Canadian industry, academia, Government of Canada organizations such as PHAC, and other international space agencies and organizations.

The CSA and several other space agencies have made important investments in infrastructure over the past decades in order to provide continuous observations of the Earth from space. The smart use of satellite data to develop solutions to key challenges on Earth and in our everyday lives is now being applied to multiple sectors.

This book is a step forward in the development of new solutions and applications in the public health sector and the investigation of several priority themes to which EO and geomatics tools can make important contributions: mosquito-borne and tick-borne diseases; water-borne diseases; air quality and extreme heat effects; geospatial indicators of vulnerable human populations; and pandemics.

I trust that this book can be helpful to public health decision makers as well as to the general public by providing new opportunities and ideas to solve ongoing and future public health challenges using EO.

I would like to express my deepest appreciation to all those who contributed to this book.

Éric Laliberté
Director General, Space Utilization
Canadian Space Agency

Acknowledgments

This book was co-led by the Public Health Agency of Canada and the Canadian Space Agency.

The authors would like to thank those who have contributed directly or indirectly to the success of this book.

Special thanks to Dirk Werle from Ærde Environmental Research, Halifax, Nova Scotia, Canada, and Kaarla Sundström from Strömcomm, Canada, for their editing services and support in the development of the manuscript.

Public Health Agency of Canada: Antoinette Ludwig, Marie-Josée Champagne, Philippe Berthiaume, Catherine Bouchard, Thomas Davis, Serge Olivier Kotchi, Marie-Laure LeCarre, Julie Légaré, Yann Pelcat, Patricia Turgeon, Erin Rees, Guillaume Poliquin, and Steven Guercio.

Canadian Space Agency: Éric Laliberté, Marie-Josée Bourassa, Martin Bergeron, Paul Briand, Steve Iris, Mike Kirby, Laurent Giugni, Patrick Plourde, Ian Hobler, and Maximilian Bonito.

Groupe de recherche en épidémiologie des zoonoses et santé publique (GREZOSP) de l'Université de Montréal, Canada: Miarisoa Rindra Rakotoarinia, Marion Ripoche, and Ludivine Taieb.

World Health Organization (WHO): Ramesha S. Krishnamurthy.

National Aeronautics and Space Administration (NASA), USA: John Haynes, Shobhana Gupta, and Helena Chapman.

National Oceanic and Atmospheric Administration (NOAA), USA: Julie Trtanj.

Centre National d'Études Spatiales (CNES), France: Cécile Vignolles.

Group on Earth Observations (GEO): Steven Ramage and Mike Gill (GEO BON).

Institute of Research for Development Espace Dev (IRD), France: Thibault Catry.

Institut National de Santé Publique du Québec (INSPQ), Canada: Marie-Ève Turcotte.

Environnement and Climate Change Canada: Philippe Martin.

Natural Resources Canada: André Beaudoin.

Statistics Canada: Marcelle Grenier.

Université de Sherbrooke, Canada: Richard Fournier.

University of Vienna, Austria: Marion Borderon.

Université du Québec à Montréal (UQAM), Canada: Yves Baudouin.

Université de Montréal, Canada: François Cavayas.

Université Laval, Canada: Gabriel Ahui.

Université du Québec à Rimouski (UQAR), Canada: Claudia Carrascal.

Arctus Inc., Canada: Thomas Jaegler.

Queen's University, Canada: Dongmei Chen.

Eureka Géo-Consultants, Canada: Yuddy Ramos.

Quadram Institute, UK: Matthew Gilmour.

Executive Summary

This document is the result of extensive information gathering on the potential for Earth Observation (EO) to contribute to public health practice. The information was primarily provided by remote sensing experts from the EO community together with epidemiologists, modeling experts, policy makers, managers, and public health researchers who gathered at the One Earth – One Health workshop held at the Canadian Earth Observation Summit in Montreal in 2017.[1] At this event they shared how EO is being used to understand, track, predict, and manage infectious diseases and discussed the challenges and significant potential of using and developing EO data for public health purposes. The information provided by these workshop participants, and beyond (2017–2021) along with other members of the international community, has been compiled in this book to reach a greater number of EO community members and public health professionals interested in developing and applying EO and other geospatial applications in the risk assessment and management of public health issues.

The main objectives of this book are to answer the questions: How does EO currently assist public health activities? What are the challenges for operational use of EO in public health? What opportunities are there to further develop EO for the future benefit of public health? This book examines several priority themes to which EO and geomatics can make important contributions: mosquito-borne and tick-borne diseases; water-borne diseases; air quality and extreme heat effects; geospatial indicators of vulnerable human populations. The contributions and potential of EO as a tool to explore indicators of the COVID-19 pandemic, such as risk due to air quality and rapid environmental and socio-economic changes and impacts are also examined.

As examples presented in this book reveal, the risk of infectious disease emergence increases with a wide range of conditions and variables, including those associated with humans, animals, climate, and the environment. The growing awareness of complex but causal interactions among these realms has motivated professionals in a wide range of sectors to adopt the One Health approach, which promotes inter-sectoral collaboration to address health issues at the human–animal–environment interface.[2] In its 2030 Agenda for Sustainable Development, the United Nations specifically identifies "strengthening the capacity of all countries, in particular developing countries, for early warning, risk reduction and management of national and global health risks" as part of their Good Health and Well-being Sustainable Development Goal (SDG).[3] In the context of both the One Health concept and the SDG initiative, remote sensing can provide solutions to the priority of assessing and monitoring public health risks, and it can play an important role in supporting decision making to reduce health risks within our shared ecosystems.

Remote sensing provides the detailed EO data that are particularly useful for risk modeling and mapping projects, which in turn generate information on the occurrence and spatio-temporal trends in disease risk. EO data can be used in early warning systems to identify risk factors that can lead to the emergence of a disease. Risk maps enable public health professionals to anticipate and prepare for health threats, and they can support responses to infectious disease epidemics or existing endemic conditions such as non-infectious chronic diseases. By identifying where disease risks are most likely occurring, risk maps help public health manage and mitigate threats to health. The maps may identify risks before they emerge, giving public health experts the time to develop policies to prepare, respond to, and build resilience to health threats. They can guide the design of surveillance programs and help support decision making in outbreak situations, including implementation of interventions for the prevention and control of disease risks.

While EO has generally proven itself in identifying environmental determinants of disease hazards, it also offers the potential to characterize and locate vulnerable populations. EO data and information products contain useful socio-economic indicators, such as the characterization of built environments and human population density. Geospatial information on environmental hazards (i.e. the exposure of the population to diseases) and human population vulnerability indicators (including sensitivity to disease and capacity to respond) can be combined to produce true risk maps that illustrate environmental hazards and population vulnerability. Thus, public health decisions and actions can be more efficiently targeted to specific populations – for example, the elderly or young children – in risk-prone environments.

This book explores the use of EO data using case studies that involve risk mapping for mosquito-borne and tick-borne infectious diseases and monitoring for applications related to air quality and water quality. Practitioners point to the fact that collecting relevant EO data and generating thematic information products requires significant know-how, infrastructure, and partnerships for data management and process automation. The resulting products address the vital need for open access and up-to-date, accurate, and authoritative data for evidence-based decision making in public health. Local public health-related data and information requirements could also be supported by existing global EO data collection infrastructure, services and thematic products serving security, marine-environmental, atmospheric, and land cover and land use applications. There is also a need for closer collaboration between space agencies, organizations responsible for EO data generation, acquisition, and distribution, and the public health community in order to obtain and utilize EO data that effectively address health issues. Epidemiological data need to be collected and accessed to complement and verify public health-related geospatial data and information products, which are often obtained by the use of surveillance data for calibration and validation of predictive models.

The public health community faces challenges using, accessing, and maintaining access to timely, reliable, and accurate EO data, while at the same time the need for high-volume health-related data on environmental, climatic, and socio-economic factors is increasing domestically and internationally. The availability of the technical infrastructure to support the use of EO data and its transformation into products useful for public health-related analyses and decision making is an essential prerequisite.

Appropriate EO tools and products, as well as interoperable EO-based products need to be developed, and spatial, analytical, and timely solutions need to be established in partnership with the public health community. Advances in artificial intelligence and big data management promise analysis-ready data for risk models to create timely risk maps.

Ultimately, the usefulness of EO satellites depends in large part on the ability of users to access and apply the data and technology in practical settings in order to address pressing issues. More could be accomplished with the participation of users and science teams with remote sensing expertise from different sectors, including public health, in mission planning and establishing utilization cycles. In addition, a collaborative, multi-sector approach would be useful in identifying observation requirements, specifying technical needs, and designing new instruments to meet a wide range of requirements and EO data needs. Being able to influence missions and produce data and algorithms specific to monitoring and managing health issues would be another way for public health to benefit more fully from remote sensing. However, involvement in this type of mission development requires skilled human resources on multiple levels.

The many needs, opportunities, and solutions described in this book would all benefit from a more extensive Community of Practice committed to using and improving how geospatial information is adopted and managed for public health issues. And just as monitoring and managing our planet involve understanding the complex interactions of natural forces, ecosystems, and anthropogenic activities around the world, so too must our responses to the challenges of more effectively using EO in public health involve multi-sectoral, cross-border cooperation. Collaboration is at the foundation of innovative action and the need for action on infectious and chronic diseases is only increasing with globalization, climate change, and natural and human-induced stresses on and in the environment.

The following is a brief description of the six themes that are the focus of this book.

Mosquito-borne Diseases

- Identifying and monitoring vector populations has been highlighted by the World Health Organization as an important component of its global surveillance efforts.
- Mosquitoes have a short life cycle (as short as a few weeks). Using a combination of EO technology approaches, EO could play a crucial role in identifying risk locations for mosquito-borne diseases locally and regionally on the basis of habitat and climate variables when these data have high to very high spatial and temporal resolution (Appendix B).
- EO could support current research efforts to develop weather-based and environment-based statistical forecasting of places and times at which risk (or hazard) of mosquito-borne diseases (e.g. malaria and Rift Valley fever) is high.

Tick-borne Diseases

- Lyme disease is the most reported vector-borne disease in the northern temperate climatic zone, occurring in North America, Europe, and Asia. There is evidence that climate change is driving the spread of ticks in Canada, and the geographic range of endemic Lyme disease risk in Canada is expected to expand as the climate warms.
- Ticks have multi-year life cycles, and EO can assist in the risk modeling of occurrence of ticks such as *Ixodes scapularis* – an important vector of Lyme disease – because it can identify climate and woodland habitats that are suitable for ticks and their hosts.
- Observations from EO can be used in various ways to contribute to risk assessments, especially when collected over areas where ground observations are limited, absent, or unavailable. Currently available EO, integrated to create maps of land cover and used at 30 m spatial resolutions and finer, will allow Lyme disease risk areas to be identified with more detail than previously possible.

Air Quality and Heat

- Epidemiological evidence suggests that air pollution has become the biggest environmental cause of premature death, overtaking poor sanitation and a lack of clean drinking water. Nine out of 10 people breathe air containing high levels of pollutants; it is estimated that around seven million people die every year from exposure to fine particles in polluted air, which

can cause stroke, heart disease, lung cancer, chronic obstructive pulmonary diseases, and respiratory infections, including pneumonia.
- EO data routinely support monitoring of heat and air quality by a variety of measurement and mapping activities, including estimating emissions, tracking pollutant plumes, supporting air quality forecasting, detecting wildfires, providing evidence for exceptional event declarations, monitoring regional long-term trends, and evaluating air quality model outputs.
- Operational products are now available in North America to forecast air quality, including when air quality issues are associated with wildfires. Effective risk models, critical support for operation, and significant infrastructure with high-performance computing systems are required to process EO data on a daily basis and to generate operational products.

Water-borne Diseases

- Non-cholera *Vibrio* (NCV) bacterial infections in humans can cause mild, self-limiting gastroenteritis, but they can also infect wounds, which can rapidly result in septicemia and necrotizing fasciitis with a high fatality rate. Climate change is expected to disproportionally affect NCV risk in coastal regions in northern and southern latitudes and is likely to greatly expand the geographic range of NCV.
- There is a lack of data on the epidemiology of NCV and a lack of data on NCV in the environment with sufficient resolution in time and space that can be utilized for public health practice. Thus, EO services can be used as proxies for this data deficit and can monitor the environmental precursors of this disease.
- The active Vibrio Map Viewer developed by the European Centre for Disease Control (ECDC) shows that sea surface temperature forecasts from modeling using EO data can be used as climatic indicators of *Vibrio* bacteria growth in coastal regions of the Baltic Sea as part of an early warning system for NCV. These types of EO systems for coastal monitoring are intended to reduce the burden of disease in human populations by providing timely warnings to public health organizations.

Vulnerable Populations

- Human vulnerability is defined by the complex interaction between the susceptibility of an individual, community, or population exposed to a threat and their capacity to reduce the risk or impact. Assessing human vulnerability to infectious disease is important in order to target populations most at risk and to reduce the burden of illness.
- EO can provide baseline information on habitable vs. non-habitable areas for a variety of environments and on the location of a population, including rural, urban, and even hard-to-reach "invisible" populations such as dislocated and migratory peoples.
- Innovative methods have combined socio-economic and demographic EO data proxies to produce public health information, such as the example of the malaria infection risk index.
- A vulnerability index could be integrated into many health studies and models to identify populations at risk.

Pandemics

- The COVID-19 pandemic has stimulated innovative development. EO players, satellite operators, and the Group on Earth Observations community of public health practice have been able to rely on regular EO satellite operations and data delivery services that functioned nominally and were uninterrupted during the pandemic.

- EO applications have proven their utility in: assessing critical environmental conditions (e.g. air quality, land use, weather); tracking compliance levels with regard to public health measures (e.g. mobility, activity levels day and night); monitoring and supporting certain COVID-19 interventions (e.g. lockdowns, vaccines); and providing data where civilian authorities need support to target vulnerable populations and estimate population density and mobility.
- EO can deliver fast and detailed information about air quality that can provide important data for understanding respiratory pandemics like COVID-19 and forecasting severity of symptoms in those infected.
- Satellite imagery can help civilian authorities plan for pandemic response and recovery phases. Generating actionable knowledge and public information regarding the dynamics of the COVID-19 pandemic will be an urgent task over the next few years. Research concerning key EO data collections and curatorial services will be essential for providing evidence-based data, information, and know-how for decision makers in the public health Community of Practice.

Notes

[1] https://crss-sct.ca/conferences/csrs2017/one-earth-one-health-workshop/ (accessed 6 January 2022).
[2] https://www.who.int/features/qa/one-health/en/ (accessed 6 January 2022).
[3] https://www.un.org/development/desa/disabilities/envision2030.html (accessed 6 January 2022).

1 Introduction to Public Health and Earth Observation

1.1 Public Health and Earth Observation

Infectious and chronic diseases are issues of concern for public health on a global, regional, and local level. Key to managing these diseases and reducing their impact is having timely, evidence-based knowledge. Earth Observation (EO) provides data at multiple spatial scales and is becoming a vital tool in helping us understand, track, and predict these diseases, allowing public health to proactively plan and implement informed interventions. This book will illustrate current and possible future contributions of EO to public health practice.

Most emerging infectious diseases of significance to public health originate in wildlife and then spill over into human populations. Research has led to improved detection and control of infectious diseases and has expanded our knowledge of how these diseases emerge and re-emerge as driven by a combination of factors that include genetic change in causal pathogens, climate and other environmental changes, and changing human behavior. Emerging infectious diseases pose continuous challenges to public health preparedness and policies and to programs for surveillance, prevention, and control (Jones *et al.*, 2008).

The increasing risk of disease emergence, epidemics, and pandemics has been documented, even before the COVID-19 pandemic swept across the globe. Worldwide, infectious diseases are responsible for 14 million deaths every year. More than 90% of these deaths occur in low- and middle-income countries in the Global South, where infectious diseases account for 43% of all causes of death versus only 1% in high-income countries in the Global North (Sénat, 2012). However, the incidence of emerging diseases in high-income countries has risen from 10 to 20% over the first decade of the 21st century, and a total of 335 new infectious diseases were discovered between 1940 and 2004. Infectious diseases create serious economic barriers to global development due in part to their association with increasing societal and financial costs (Sénat, 2012). Approximately two out of three human infectious diseases are zoonotic, meaning they are (or were originally) animal diseases that are transmitted to humans. In addition, at least three out of four emerging infectious diseases among the human population are or were zoonoses, and on average five new diseases appear every year (World Organisation for Animal Health [OIE], World Health Organization [WHO]).[1]

Global environmental change and biodiversity loss are exerting major pressure on human health (United Nations Environment Programme [UNEP] *et al.*, 2015; UNEP, 2020). According to the WHO's fact sheets[2]: rabies transmitted by vampire bats to cattle and humans has

DOI:10.1079/9781800621183.0001

been linked to forest activities in South America; the spread of Nipah virus has been linked to intensification of pig farming and fruit production in Malaysia; the emergence of Japanese encephalitis virus has been linked to irrigated rice production and pig farming; and the emergence of avian influenza has been linked to intensive poultry farming (WHO, 2018). The cholera bacteria transmitted in water and the dengue virus and malaria parasites transmitted by mosquitoes infect 3–4 billion people every year, and outbreaks of disease associated with these pathogens are often driven by environmental factors (WHO, 2021a, 2021b).

Key drivers of infectious diseases and the One Health approach

Infectious diseases emerge and re-emerge under the influence of key drivers. By understanding how these drivers affect diseases, we may be able to predict when, how, and where disease will emerge and to identify the populations that are most at risk. Examples of drivers include environmental, climatic, demographic, socio-economic, or human behavioral changes. While "risk" is truly the combination of rates of exposure to a "hazard" (e.g. a vector-borne disease) and the "susceptibility/sensitivity" of a population to that hazard, in this book, the term "risk" is sometimes used instead of "hazard" when this has been commonplace in the literature.

The world's human population – presently exceeding 7.5 billion people – is expected to reach 9.7 billion by 2050. Many people will continue to concentrate in megacities and large metropolitan areas, which facilitates human-to-human disease transmission (Neiderud, 2015). Ecosystem changes in land use and agricultural practices, such as deforestation, intensive livestock farming, and the movement of animals between forests and cities will likely increase people's exposure to wildlife-borne diseases directly or indirectly through infected livestock (Jones *et al.*, 2013).

Global environmental change, including climate change, is accelerating species loss, leading toward a biodiversity crisis, and this loss in biodiversity is associated with the emergence of infectious disease (Keesing *et al.*, 2010; Ostfeld and Keesing, 2012; Altizer *et al.*, 2013). Global increase of trade in goods and animals can also contribute to the spread of disease vectors (Tatem *et al.*, 2006). Increased air transport accelerates the movement of people into and out of risk areas, and "naïve" populations in countries free from a particular disease are increasingly threatened by infected tourists and business people returning from countries where the disease is endemic (WHO[3]). Climate change is likely to change the geographic range where climate is favorable for multiplication of arthropod disease vectors, such as mosquitoes and ticks. Examples include the observed expansion of Lyme disease in northern North America and the possible expansion of risk from dengue and chikungunya into regions that were previously temperate (Ogden, 2017). Displacement of populations as a result of natural disasters, scarcity of water resources, famine, or wars is confronting us with new diseases as people move into new geographical areas (WHO, 2006). Also, resistance to antibiotics and increase in virulence of pathogens may drive disease emergence (WHO[4]; Beceiro *et al.*, 2013).

Given the close and complex relationships between the environment, ecosystems, and the etiological agents of disease in human and animal populations, integrated approaches following the One Health approach[5] are most likely to be successful as they take into consideration human, animal, and environmental health with interdisciplinary collaborations and communication in all aspects of health. In this context, many parameters and geospatial characteristics relevant to the interconnected fields of environmental, human, and animal health can be assessed via proxy measurements from space. However, efficient methods and pertinent One Health partnerships need to be developed in order to adopt satellite-based remote sensing as a suitable tool for characterizing, mapping, and monitoring risk factors for infectious disease emergence and re-emergence. Partnerships between the EO community and the public health community would be a first step toward this goal.

Use of EO data in public health practice

EO data have proven to be a valuable source of geospatial information for public health, particularly in the realm of "tele-epidemiology."

Based on EO products adapted to the needs of health actors, tele-epidemiology studies the links between the environment, ecosystems, and etiological agents responsible for diseases in human, animal, and plant populations. This approach combines the physical, biological, social sciences, and humanities, and aims to understand the factors and mechanisms that affect the spread of pathogens and diseases (Marechal *et al.*, 2008). Environmentally linked diseases, including vector-borne, water-borne, and airborne diseases, have geographic distributions at global, national, regional, local, and neighborhood scales that are associated with the geographic distributions of the climatic, habitat, and land use factors that determine their transmission (Eisen *et al.*, 2015; Kilpatrick *et al.*, 2017). Surveillance is the gold standard method of identifying disease risk (Ogden *et al.*, 2014; Bouchard *et al.*, 2015), but the vastness of the Earth renders surveillance at every location all but impossible. Consequently, for environmentally linked diseases, point data obtained in surveillance are increasingly being used to calibrate and validate models that identify associations between environmental variables and environmentally linked disease occurrence. These associations can then be used to extrapolate occurrence of risk onto surfaces to create risk maps (e.g. Soucy *et al.*, 2018), provided the environmental variables are present as a continuous surface. The continuous surfaces of EO data proxies for environmental variables, which have the same precision across the globe, are a significant reason why they are so useful for creating risk maps of disease emergence and spread (Michel *et al.*, 2011; Cheng *et al.*, 2017).

EO data furnish the development of proxies for environmental drivers of diseases, such as habitat (e.g. forest type and density, presence of wetlands), agricultural areas and types, surface temperature, soil moisture, and urban areas. With the recent improvement of satellite EO systems, it is now possible to increase observations and monitoring of land and water parameters (i.e. weather, climate, population distribution, animal habitat identification, etc.) in repeated, synoptic, low- to large-scale ways. These recent innovations increase spatio-temporal precision of EO data and offer the possibility of improved model-based identification of risk in public health research. Greater spatial precision allows more detailed risk maps to be produced, while greater temporal precision (i.e. near-real-time EO data) raises the possibility that EO data proxies for weather may facilitate disease forecasting (e.g. Ogden *et al.*, 2019) and EO data may be used to assist on-the-ground activities in response to outbreaks. As EO data can provide proxy measurements for socio-economic factors that may be determinants regarding the sensitivity and adaptive response capacity of the human population, EO data can in theory measure and communicate all aspects of disease risk.

Mandate and role of public health organizations and the importance of geospatial information

In order to address public health issues with relevant geospatial approaches, technology for detailed data collection is an essential component of fulfilling the mandate and role of public health organizations. Public health encompasses the organized efforts of society to keep people healthy and to prevent illness, injury, and premature death (Feinleib, 2001). In the Canadian context, the Public Health Agency of Canada (PHAC), in collaboration with all three levels of government, the private sector, non-governmental organizations, health professionals, and the public, contributes to: the prevention of disease and injury; the promotion of health; and sharing public health expertise across Canada and with international partners. Below is a summary of related activities (from the PHAC mandate[6] and roles; 2021–2022 PHAC report on Departmental Plans[7]). PHAC has the responsibility to:

- Contribute to the prevention of disease and injury, and to the promotion of health.
- Strengthen intergovernmental collaboration on public health and facilitate national approaches to public health policy and planning.
- Provide federal leadership and accountability in managing national public health events.
- Enhance surveillance information and expand the knowledge of disease and injury in Canada.
- Serve as a central point for sharing public health expertise across Canada and with

international partners, and to use this knowledge to inform and support Canada's public health priorities.

Through PHAC's research, programs and services, its goals are to bring about healthier Canadians, reduced health disparities, and a stronger capacity to deliver on and support public health activities.[8]

PHAC promotes key initiatives and activities that provide Canadians and public health stakeholders with the science, research, guidance, and resources to build public health capacity across Canada. To deliver its mandate, PHAC must be able to support a variety of activities (from the PHAC Mandate and Role, Corporate Risk Profile and Departmental Plan) in order to:

- Provide meaningful data and information with technology and experimentation, allowing for more timely identification of public health issues and the development of novel, evidence-based solutions to address them.
- Apply international research and development to Canada's public health programs, focusing on key initiatives and activities that provide Canadians and public health stakeholders with the science, research, guidance, and resources for infectious disease prevention, and increase awareness, while reducing harms.
- Prepare for, and respond to public health emergencies, including infectious diseases, by predicting, detecting, assessing, and responding to outbreaks and new threats, and contribute to the prevention, control, and reduction of the spread of infectious disease among Canadians.
- Continue to focus on providing the latest data, evidence, and scientific information required to respond to both ongoing infectious diseases and new outbreaks.
- Strengthen intergovernmental collaboration on public health and facilitate national approaches to public health policy and planning.

Within PHAC, the National Microbiology Laboratory (NML) conducts research, laboratory diagnosis, risk assessments, and laboratory-based surveillance for emerging infectious diseases such as COVID-19, Lyme disease, West Nile virus, Zika virus, Ebola hemorrhagic fever etc.[9]

The Public Health Risk Science division (PHRS) of the NML works to understand where and when risks from infectious diseases occur, and how best to prevent and control them. This includes diseases within the food chain and the environment and those that arise from contact between humans and animals.

This division:

- undertakes epidemiological studies to identify public health risk factors;
- develops risk models and decision analysis tools to understand and reduce public health risks; and
- develops public health geomatics (geographical information) tools and services to support decision making by emergency response teams and surveillance programs.

PHRS undertakes research to generate and disseminate disease risk assessment, information, and tools for prevention and control. In developing risk assessments, the division takes a transdisciplinary One Health approach, using mathematical modeling, epidemiology, remote sensing, and geomatics, supported by knowledge synthesis capacity. These activities support programs in public health surveillance, research, and outbreak management.[10]

PHAC supports surveillance and monitoring of infectious diseases, risk assessments, modeling, and laboratory diagnostics. Now and in the future, PHRS aims to strengthen its ability to assess the risk of diseases and to improve detection, monitoring, and prediction of existing and emerging infectious disease threats through the continued implementation of new technologies. EO satellite technologies are relatively new in the public health context so their development and application in public health have so far been done in close collaboration with their expert partners at the Canadian Space Agency (CSA).[11]

Mandate and role of space agencies and the importance of EO

National and international space agencies offer a range of assets, programs, infrastructure, and expertise to improve surveillance, emergency response and preparedness, and effective early-warning and information systems. EO data and remote sensing know-how are key elements for public health agencies in their efforts to detect environmental variables that influence the emergence and spread of diseases. In Canada, the CSA and its EO application programs and activities facilitate the development of innovative solutions and activities, including strengthening quick-response capacities to counter public health threats with EO.

The CSA's raison d'être is to lead the development and application of space knowledge for the benefit of Canadians and for all of humanity.[12] Core responsibilities of the CSA include the following:

- Coordinating space policies and programs of the government of Canada.
- Ensuring access to space data, information, and services for other government departments and agencies to deliver on their mandates.
- Planning, directing, and managing projects related to scientific or industrial space research and the development of space science and technology.
- Promoting the transfer and diffusion of space technology to and throughout Canadian industry.
- Encouraging the commercial exploitation of space capabilities, technology, facilities, and systems.

The CSA also aims to build Canada's capacity to engage the next generation of space scientists and engineers, and provide opportunities to inspire young people to develop the required skills and to pursue studies and careers in science, technology, engineering, and math.[13] The mandate of the CSA is "to promote the peaceful use and development of space, to advance the knowledge of space through science and to ensure that space science and technology provide social and economic benefits for Canadians. The CSA is delivering on its mandate in collaboration with Canadian industry, academia, Government of Canada organizations, and other international space agencies and organizations."[14]

Collaborative EO efforts are formalized under national and international partnership agreements, memorandums of understanding (MOU), and contracts. CSA is focused on ensuring that: (i) space research and development is advanced in terms of science and technology; (ii) Canadians are engaged with space; (iii) space information and technologies improve the lives of Canadians; and (iv) Canada's investments in space benefit the Canadian economy.[15] In the area of EO, research and development funding support from CSA has helped other government departments to advance scientific research and reach out into operation domains. Much effort has been devoted to developing cost-effective use of synthetic aperture radar satellite data.

Following the success of the RADARSAT-1 and RADARSAT-2 satellite EO missions, the CSA launched the RADARSAT Constellation Mission (RCM) on 12 June 2019. The three identical radar satellites are operating independently of weather and light conditions. They acquire detailed EO data in a coordinated way to address key Canadian challenges, including public health-related issues. The RCM orbital configuration allows for daily revisits of Canada's vast territory and maritime approaches and exact revisits every 4 days, as well as access to 90% of the world's surface every day and the Arctic up to four times a day. Over a dozen Canadian federal government departments, including PHAC, already use RADARSAT data to deliver important services to Canadians. RCM will ensure the ongoing availability of these data so that the Government of Canada can continue to serve Canadians. RCM is designed to provide effective solutions in three main areas: maritime surveillance (ice, surface wind, oil pollution, ship monitoring); disaster management (mitigation, warning, response, recovery); and ecosystem monitoring (agriculture, wetlands, forestry, coastal change monitoring). Public health organizations and authorities will benefit from the new RCM data collection as well as the archival data sets.

The CSA's EO programs and activities include funding opportunities for academic research. They help to increase industry capabilities, and support research and development activities in government organizations. Benefits include:

- Improvements in EO satellite data availability.
- EO-derived information related to land cover and use, climate, population distribution, and identification of habitats.
- Advances in the spatial-temporal resolution and quality of geospatial data.
- Integration of EO-derived data sets and related geospatial information into predictive

models, surveillance programs, and emergency management activities with important domestic and international dimensions.

Partnership of the PHAC and the CSA

Over the past decade, the collaboration between PHAC and CSA has developed into a successful partnership to advance the application of space technologies and geospatial data in public health. Both organizations have actively participated in international committees and a series of domestic research and development projects concerning the prevention and control of infectious diseases.

Some examples of their successful joint activities include the collaborative projects of PHAC, CSA, and other Canadian government departments on risk assessment for microbial contamination of recreational waters using satellite imagery. Joint projects with industry partners, in support of One Health initiatives, used RADARSAT data to study water detection and to monitor wetlands and lake extent to support efforts to identify water-borne and mosquito-borne disease risks. There have also been joint activities with academia focused on health-related projects in urban environments and research into public health threats such as heat, air pollution, and mosquito-borne diseases that take into account the impact on vulnerable human populations. The projects were supported through the CSA Government Related Initiatives Program (GRIP), the Earth Observation Applications Development Program (EOADP), and the Science and Operational Applications Research (SOAR).

At the international level, PHAC and CSA participated from 2006 to 2015 in the United Nations Convention on the Peaceful Use of Outer Space (UN-COPUOS) Action Team 6 on Public Health. The mandate of this Action Team involved the implementation of telehealth plans and activities to improve health services in developing countries by facilitating the application of space technologies in early warning of infectious diseases. Since 2015, PHAC and CSA have been participating in the newly formed Expert Group on Space and Global Health. This group engages Member States and international governmental and non-governmental organizations in collaborative projects and is tasked to propose tangible and long-lasting solutions regarding the contribution of space to the global health agenda. PHAC and CSA contributions have been documented in part in several United Nations reports. They include a Special Report of the Inter-Agency Meeting on Outer Space Activities on the use of space science and technology within the United Nations system for global health[16]; a Report on the Meeting on the Applications of Space Science and Technology for Public Health organized by the World Health Organization and the Office for Outer Space Affairs[17]; and a Report on the United Nations Expert Meeting on the International Space Station Benefits for Health.[18] The CSA has supported a special study on tele-epidemiology in close collaboration with PHAC to better understand this emerging EO sector.

In addition to their contributions at the United Nations, PHAC and CSA have co-led international conference sessions and workshops (e.g. European Space Agency Living Planet Symposium – Special Session on Tele-epidemiology, Prague 2016; EO Summit, One Earth – One Health, Montreal, 2017) supporting the application of tele-epidemiology in the public health domain. One of the main goals has been to develop and maintain a community of practice with a focus on public health and EO through a number of activities:

- Convene leaders and experts in EO and public health to explore, discuss, establish, or strengthen collaborations and partnerships on novel EO applications, products, and services to support public health.
- Better understand the links between environment, climate, society, and public health that can be elucidated using EO data.
- Identify existing public health applications derived from EO data.
- Identify existing or potential future EO data, indicators, methods, and technologies that may be developed to support public health.

The One Earth – One Health workshop provided a forum for scenario-based discussion and dialogue among recognized experts and authorities on EO technology, applications, and methods that are relevant to application in public health. The main goal was – in alignment with the UN

Sustainable Development Goals – to explore the use of EO data to ensure healthy lives and to promote well-being for all ages.[19] The scenario-based discussions focused on how scientific and institutional cooperation with respect to promising EO applications can be developed or expanded to better serve and protect the public. The outcomes generated by this workshop form the basis of this book.

1.2 Work Program Organization and Management Approaches

The practical benefits resulting from the use of EO data streams are evident. Major scientific breakthroughs and subsequent application development work in particular have supported climate observation, resource assessments, and environmental monitoring. EO can also provide some of the geospatial information required for public health research, risk assessment, and application in programs and operations. In Canada, this has required the organization and management of a joint work program by PHAC and CSA to obtain expert knowledge to guide further development of EO applications for use in public health. The main elements of this work program are outlined in the following sections.

Identification and elaboration of key themes

At the outset, PHAC and CSA researchers and managers defined a number of key themes in collaboration with subject experts who formed an organizing committee prior to an international workshop on the theme of One Earth – One Health. The members of this committee included representatives from the Centre national d'étude spatial (CNES), the Institut de Recherche pour le Développement (IRD), Espace-Dev and VetAgro Sup Campus in France, the Université de Sherbrooke in Canada, the international Committee on Earth Observation Satellites (CEOS) Working Group on Capacity Building & Data Democracy (WGCapD), Ærde Environmental Research in Canada, and PHAC and CSA. The committee identified six themes that were to be explored in more detail during a workshop with international experts:

- mosquito-borne diseases;
- tick-borne diseases;
- air quality and chronic diseases;
- water-borne diseases;
- vulnerable human populations; and
- pandemics.

There was no priority assigned to the themes: they all relate to significant public health issues to which EO data and derived information can potentially contribute useful information for public health actions. Detailed discussions of these six themes are described in Section 2 of this book. The description includes a review of relevant literature, a brief overview of recent projects, and a scenario-based framework of inquiry. Representatives of national and international space- and public health-related organizations contributed to the development of each scenario.

In the workshop, a series of questions was posed to experts on each of the six themes. The questions aimed to identify the scope and outcome of current work; challenges and issues to be addressed during research and application development; examples of previous work; and to define potential future uses in public health.

Expert consultation process

The engagement of national and international expertise was a central element in the PHAC/CSA-led consultation process. Expert opinion was collected prior to and during the proceedings of the workshop. Important session outcomes and scenario-based discussions were synthesized for this book. Experts contributed examples and illustrations of their work to the book and critically reviewed the synthesis of workshop results and recommendations.

Based on the results of previous studies and expert consultations, CSA and PHAC assessed the research and development status and operational readiness levels of EO technology applications regarding key public health themes. The assessments emphasized the utility of EO-derived information for public health, with a view toward implementing EO data products in regular surveillance, prevention, control, prediction/ risk assessment, disease forecasting, and public outreach.

and public health-related objectives of forming new partnerships and initiatives that can support public health from local to global levels.

Identification of needs and opportunities

The identification of EO and public health-related needs and opportunities resulted from an analysis of the relevant scientific literature and subject-specific presentations of the experts during the One Earth – One Health workshop. They reflect current activities and goals. The various needs and opportunities were subdivided into eight categories. Detailed in Section 3, these categories are: (i) aligning with and supporting UN Sustainable Development Goals; (ii) focusing on public health needs and key theme areas for further research; (iii) accessing and developing EO and geospatial evidence-based data/products leveraging public health capacities; (iv) developing a sustainable community of practice; (v) developing knowledge and know-how; (vi) developing solutions: methods, tools, and systems; (vii) implementing technical infrastructures and technologies; and (viii) participating in EO satellite mission development for monitoring disease risks. These categories serve as a guide for further action to achieve specific EO and public health-related objectives of forming new partnerships and initiatives that can support public health from local to global levels.

Objectives and book outline

This book addresses three basic questions: How does, or can, the current capacities of EO assist public health activities? What are the challenges for the operational use of EO in public health? And what opportunities are there to further develop EO to benefit public health in the future? To answer these questions, this book identifies key public health activities in which EO data are or can be used. This includes prediction of disease emergence and spread and of disease forecasting to support public health programs for disease surveillance, prevention, and control interventions. More specifically, the book aims to: (i) assess current research and identify and document key themes; (ii) collate expert advice from the Canadian and international EO and public health communities on specific themes; and (iii) present conclusions and opportunities. Our goal is to guide decision-making on further research and on the development of innovative EO applications and solutions in the public health sector.

Notes

[1] https://www.oie.int/en/what-we-do/global-initiatives/one-health/ (accessed 22 December 2021), http://www.emro.who.int/fr/about-who/rc61/zoonotic-diseases.html (accessed 22 December 2021).

[2] https://www.who.int/news-room/fact-sheets (accessed 22 December 2021).

[3] https://www.who.int/news-room/q-a-detail/health-risks-when-traveling (accessed 22 December 2021).

[4] https://www.who.int/news-room/fact-sheets/detail/antimicrobial-resistance (accessed 22 December 2021).

[5] https://onehealthinitiative.com/about/ (accessed 22 December 2021).

[6] https://www.canada.ca/en/public-health/corporate/mandate/about-agency/mandate.html (accessed 18 January 2022).

[7] https://www.canada.ca/en/public-health/corporate/transparency/corporate-management-reporting/reports-plans-priorities/2020-2021-corporate-information.html (accessed 18 January 2022).

[8] https://www.canada.ca/en/public-health/corporate/mandate/about-agency/mandate.html (accessed 18 January 2022).

[9] https://www.canada.ca/en/public-health/programs/national-microbiology-laboratory-for-professionals.html; West Nile, Lyme Disease Surveillance: https://www.canada.ca/en/public-health/services/surveillance.html#a17; COVID-19 surveillance: https://www.canada.ca/en/public-health/services/diseases/2019-novel-coronavirus-infection/health-professionals/interim-guidance-surveillance-human-infection.html (all accessed 18 January 2022).

[10] https://www.canada.ca/en/public-health/programs/national-microbiology-laboratory-for-professionals.html (accessed 18 January 2022).

[11] https://www.asc-csa.gc.ca/eng/satellites/everyday-lives/how-satellites-help-you-stay-healthy.asp (accessed 18 January 2022).
[12] https://www.asc-csa.gc.ca/eng/publications/dp-raison-d-etre.asp (accessed 22 December 2021).
[13] https://www.asc-csa.gc.ca/eng/publications/dp-2019-2020.asp#results-core (accessed 22 December 2021).
[14] https://www.asc-csa.gc.ca/eng/publications/dp-raison-d-etre.asp (accessed 22 December 2021).
[15] https://www.asc-csa.gc.ca/eng/publications/dp-2019-2020.asp#results (accessed 22 December 2021).
[16] UN Document: A/AC.105/1091, 30 April 2015.
[17] UN Document: A/AC.105/1099, 29 October 2015.
[18] UN Document: A/AC.105/1069, 10 September 2014.
[19] https://sdgs.un.org/goals; http://www.earthobservations.org/sbas.php (accessed 22 December 2021).

References

Altizer, S., Ostfeld, R.S., Johnson, P.T., Kutz, S., and Harvell, C.D. (2013) Climate change and infectious diseases: From evidence to a predictive framework. *Science* 341, 514–519. DOI: https://doi.org/10.1126/science.1239401.

Beceiro, A., Tomás, M., and Bou, G. (2013) Antimicrobial resistance and virulence: A successful or deleterious association in the bacterial world? *Clinical Microbiology Reviews* 26, 185–230. DOI: https://doi.org/10.1128/CMR.00059-12.

Bouchard, C., Leonard, E., Koffi, J.K., Pelcat, Y, Peregrine A. *et al.* (2015) The increasing risk of Lyme disease in Canada. *Canadian Veterinary Journal* 56, 693–699. PMID: https://www.ncbi.nlm.nih.gov/pubmed/26130829.

Cheng, A., Chen, D., Woodstock, K., Ogden, N.H., Wu. X. *et al.* (2017) Analyzing the potential risk of climate change on Lyme disease in Eastern Ontario, Canada using time series remotely sensed temperature data and tick population modelling. *Remote Sensing* 9, 609. DOI: https://doi.org/10.3390/rs9060609.

Eisen, R., Eisen, L., Ogden, N.H., and Beard, C.B. (2015) Linkages of weather and climate with *Ixodes scapularis* and *Ixodes pacificus* (Acari: Ixodidae), enzootic transmission of *Borrelia burgdorferi*, and Lyme disease in North America. *Journal of Medical Entomology* 53, 250–261. DOI: https://doi.org/10.1093/jme/tjv199.

Feinleib, M. (2001) A Dictionary of Epidemiology, Fourth Edition. In: Last, J.M., Spasoff, R.A., and Harris, S.S. (eds.) *American Journal of Epidemiology* 154, 93–94. DOI: https://doi.org/10.1093/aje/154.1.93-a.

Jones, K.E., Patel, N.G., Levy, M.A., Storeygard, A., Balk, D. *et al.* (2008) Global trends in emerging infectious diseases. *Nature* 451, 990–993. DOI: https://doi.org/10.1038/nature06536.

Jones, B., Grace, D., Kock, R., Alonso, S., Rushton, J. *et al.* (2013) Zoonosis emergence linked to agricultural intensification and environmental change. *Proceedings of the National Academy of Sciences of the United States of America (PNAS)* 110, 8399–8404. DOI: https://doi.org/10.1073/pnas.1208059110.

Keesing, F., Belden, L.K., and Daszak, P. (2010) Impacts of biodiversity on the emergence and transmission of infectious diseases. *Nature* 468, 647–652. DOI: https://doi.org/10.1038/nature09575.

Kilpatrick, A.M., Dobson, A.D.M., Levi, T., Salkeld, D.J., Swei, A., *et al.* (2017) Lyme disease ecology in a changing world: Consensus, uncertainty, and critical gaps for improving control. *Philosophical Transactions of the Royal Society B Biological Sciences* 372, 20160117. DOI: https://doi.org/10.1098/rstb.2016.0117.

Marechal, F., Ribeiro, N., Lafaye, M., and Güell, A. (2008) Satellite imaging and vector-borne diseases: the approach of the French National Space Agency (CNES). *Geospatial Health* 3, 1–5. DOI: https://doi.org/10.4081/gh.2008.226.

Michel, P., Brazeau, S., Ogden, N.H., and Kaya, S. (2011) Integrating space technologies in public health surveillance programs: From theory to operation. *Acta Biologica Venezuelica* 31, 13–16. Available at: http://saber.ucv.ve/ojs/index.php/revista_abv/article/download/3970/3796 (accessed 5 August 2021).

Neiderud, C.-J. (2015) How urbanization affects the epidemiology of emerging infectious diseases. *Infection Ecology and Epidemiology* 5, 1. DOI: http://dx.doi.org/10.3402/iee.v5.27060.

Ogden, N.H. (2017) Climate change and vector-borne diseases of public health significance. *FEMS Microbiology Letters* 364, fnx186. DOI: https://doi.org/10.1093/femsle/fnx186.

Ogden, N.H., Koffi, J.K., Pelcat, Y., and Lindsay, L.R. (2014) Environmental risk from Lyme disease in central and eastern Canada: A summary of recent surveillance information. *Canada Communicable Disease Report* 40, 74–82. DOI: https://doi.org/10.14745/ccdr.v40i05a01.

Ogden, N.H., Lindsay LR, Ludwig A, Morse A.P., Zheng H. *et al.* (2019) Weather-based forecasting of mosquito-borne disease outbreaks in Canada. *Canada Communicable Disease Report* 45, 127–132. DOI: https://doi.org/10.14745/ccdr.v45i05a03.

Ostfeld, R.S. and Keesing, F. (2012) Effects of host diversity on infectious disease. *Annual Review of Ecology, Evolution, and Systematics* 43, 157–182. DOI: https://doi.org/10.1146/annurev-ecolsys-102710-145022.

Sénat (2012) Délégation à la prospective, Les nouvelles menaces des maladies infectieuses émergentes. Available at: https://www.senat.fr/rap/r11-638/r11-6381.pdf (accessed 4 August 2021).

Soucy, J.R., Slatculescu, A.M., Nyiraneza, C., Ogden, N.H., Leighton, P.A. *et al.* (2018) High-resolution ecological niche modelling of *Ixodes scapularis* ticks based on passive surveillance data at the northern frontier of Lyme disease emergence in North America. *Zoonoses & Public Health* 18, 235–242 DOI: https://doi.org/10.1089/vbz.2017.2234.

Tatem, A.J., Rogers, D.J., and Hay, S.I. (2006) Global transport networks and infectious disease spread. *Advances in Parasitology* 62, 293–343. DOI: https://doi.org/10.1016/S0065-308X(05)62009-X.

United Nations Environment Programme (UNEP) (2020) UNEP and biodiversity. Available at: https://www.unep.org/unep-and-biodiversity (accessed 4 August 2021).

United Nations Environment Programme (UNEP), Convention on Biological Diversity, and World Health Organization (2015) Connecting global priorities: Biodiversity and human health. Available at: https://www.cbd.int/health/SOK-biodiversity-en.pdf (accessed 5 August 2021).

World Health Organization (WHO) (2006) Disease control in humanitarian emergencies. Available at: www.who.int/diseasecontrol_emergencies/NewFlyeA3.pdf (accessed 4 August 2021).

World Health Organization (WHO) (2018) Influenza (avian and other zoonotic). Available at: https://www.who.int/news-room/fact-sheets/detail/influenza-(avian-and-other-zoonotic) (accessed 4 August 2021).

World Health Organization (WHO) (2021a) Malaria. Available at: https://www.who.int/news-room/fact-sheets/detail/malaria (accessed 4 August 2021).

World Health Organization (WHO) (2021b) Cholera. Available at: https://www.who.int/news-room/fact-sheets/detail/cholera (accessed 4 August 2021).

2 Earth Observation and Public Health Priority: Applications and Research Areas by Theme

This section presents six applications of Earth Observation (EO) to public health issues. There are also two tables in Appendix B that can guide the reader on the classes of resolution used to categorize EO systems and on EO systems and their spatial, spectral, and temporal resolution.

2.1 Mosquito-borne Diseases

Thibault Catry[1], Cécile Vignolles[2], Serge Olivier Kotchi[3], Stéphanie Brazeau[3], Antoinette Ludwig[3], Nicholas H. Ogden[3], Dominique J. Bicout[4], Richard A. Fournier[5], and Dirk Werle[6]

[1]ESPACE-DEV, Univ. Montpellier, IRD, Univ. Antilles, Univ. Guyane, Univ. Réunion, Montpellier, France; [2]Centre National d'Etudes Spatiales, Directorate for Innovation, Applications and Science, Earth Observation Programme, Toulouse, France.; [3]Public Health Risk Sciences Division, National Microbiology Laboratory, Public Health Agency of Canada, St-Hyacinthe, Canada; [4]Biomathematics & Epidemiology, EPSP, TIMC-IMAG, CNRS Grenoble Alpes University, VetAgro Sup, Veterinary campus of Lyon, Marcy l'Etoile, France; [5]Department of Applied Geomatics, Université de Sherbrooke, Québec, Canada; [6]Ærde Environmental Research, Halifax, Canada

DOI: 10.1079/9781800621183.0002

Context, state of knowledge, challenges, and responses

The World Health Organization (WHO) has highlighted identification and monitoring of vector populations as an important component of global vector-borne disease surveillance efforts.[1] (WHO, 2012, 2015) EO data could play a crucial role in identifying risk locations for mosquito-borne diseases globally on the basis of habitat and climate variables. Were EO data to have sufficiently high spatial and temporal resolution, applied research could develop weather-based and environment-based forecasting of high-risk locations and time periods for mosquito-borne diseases using statistical models. Furthermore, EO data may contribute to monitoring the evolution of risk. EO data are also useful for measuring or mapping a range of environmental parameters that help determine mosquito vector occurrence and abundance and the rate of development of mosquito-borne parasites and pathogens in mosquito vectors. These parameters include rainfall, extent of standing water, temperature, and land use and land cover.

Examples of recent research

For more than two decades, extensive research has been conducted into the use of EO data as a tool to inform responses to mosquito-borne diseases (Hay *et al.*, 1998a; Kalluri *et al.*, 2007; Kotchi *et al.*, 2019). Main objectives include identifying risk areas at various spatial scales (Rogers *et al.*, 2002), identifying seasonality in risk in different locations (Hay *et al.*, 1998b), and forecasting impending outbreaks or peaks in disease risk (Ceccato *et al.*, 2005). EO data have been used in a number of ways for these purposes. In its simplest form, EO data analysis for identifying different habitats can consist of classifying imagery into relevant landscape classes. In a case study on dengue, Machault *et al.* (2014) developed dynamic risk maps at the housing level on a daily basis for the vector mosquito *Aedes aegypti* in the French Antilles. The study identified EO data with very high spatial resolution of 0.5 m as a suitable source to produce land use classes for a spatio-temporal statistical model. Catry *et al.* (2016) fused radar and optical satellite imagery and derived land cover classifications for studying the eco-epidemiology of vector-borne diseases in tropical South America. Their study demonstrated that relevant land cover maps and wetland classifications could be generated on a weekly basis using multi-temporal cloud-penetrating C-band synthetic aperture radar (SAR) Sentinel-1A satellite data in combination with optical Sentinel-2 data and L-band SAR Advanced Land Observing Satellite-1 (ALOS) (Fig. 2.1.1).

In many parts of the world, there is insufficient ground-truthed information to reliably classify EO data as habitat that is either suitable or unsuitable for mosquito-borne disease transmission. Climate and habitat conditions must be suitable year-round for mosquito populations and pathogen transmission cycles to persist. Frequently used EO data processing techniques include ecological niche modeling, principal components analysis or Fourier processing, followed by discriminant analysis; supplemented with human case surveillance data, these techniques can be used to identify habitats that are predictive for mosquito-borne disease transmission (Rogers *et al.*, 2002; Moua *et al.*, 2021).

High levels of morbidity and mortality from mosquito-borne diseases, such as malaria, are often associated with areas where transmission of mosquito-borne diseases is unstable. This includes specific transition zones between regions where the pathogens are endemic and where environmental conditions preclude their transmission (Ewing *et al.*, 2021). The underlying reason is mostly immunological: people in transition zones are less likely to have been infected and to be immune to new infections. In researching these transition zones, EO data sets can be useful in several ways. First, they have sufficient resolution to identify these transition areas (Bejon *et al.*, 2010). Second, EO data can identify land management practices, such as irrigation, that render conditions suitable for mosquito-borne disease transmission in landscapes otherwise hostile to the vectors or transmission (Baeza *et al.*, 2013). Third, detailed EO data can identify urban environments where disease transmission may be very different from transmissions occurring in rural areas (Tatem and Hay, 2004; Ferraguti *et al.*, 2021).

While much of this research has taken place in an academic setting, there are increasing

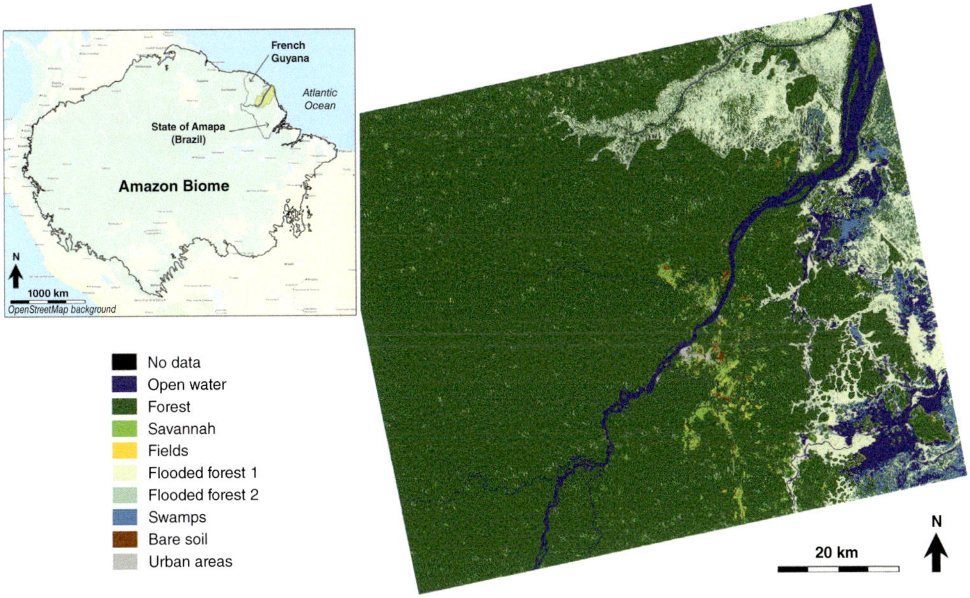

Fig. 2.1.1. Example of a land cover map based on the analysis of multi-sensor satellite imagery for classifying wetland areas in a densely forested area at the border between French Guiana and Brazil, South America. Cloud-penetrating Sentinel-1A C-band SAR data were combined with Sentinel-2 optical data (both at 10 m resolution) to produce a general land cover map. A combination of C-band and ALOS L-band SAR data was then analyzed to discriminate and map wetlands, especially flooded vegetation areas. (From: Catry *et al.*, 2016, 2018b.)

efforts to transfer knowledge gained and to implement successful EO utilization in operational mosquito-borne disease programs. An example is the MALAREO project, which has developed and implemented EO-based capabilities for national malaria control programs in the southern portion of Africa. High-resolution land cover and wetland maps were produced and integrated in a geographic information system (GIS) to identify potential vector habitats and risk associated with different human activities (Franke *et al.*, 2015). The spatial detail of the EO data has an intrinsic value for identifying and classifying habitat because ground-truthed information is rare and inconsistent. Furthermore, repeat coverage can be utilized to detect important changes with regard to habitat, land use, and land cover (Lucas *et al.*, 2015). While weather and climate may be among the most intensively measured environmental variables, interpolation of data points is a common practice in mosquito-borne disease suitability mapping. In some circumstances, EO data were found to outperform interpolated weather station data, especially in regions with a low density network of meteorological stations (Hay and Lennon, 1999).

Recent studies have shown that SAR and optical EO data are strongly complementary in the assessment of the relationships between environment components and mosquito-borne disease transmission (Machault *et al.*, 2011; Li *et al.*, 2016, 2017). EO by means of radar remote sensing has great potential to assist with the characterization of vegetated wetlands (Catry *et al.*, 2018a; see also Fig. 2.1.2). In practical and technical terms, radar capabilities are based in part on a large variety of cloud-penetrating sensors that operate at different wavelengths, polarizations, and temporal and spatial resolutions useful for wetland analyses. Furthermore, data access is facilitated by open data policies, such as those governing the use of the European Copernicus Programme and Sentinel-1 data archives. These aspects are favorable for EO research and applications regarding the epidemiology of mosquito-borne diseases like malaria.

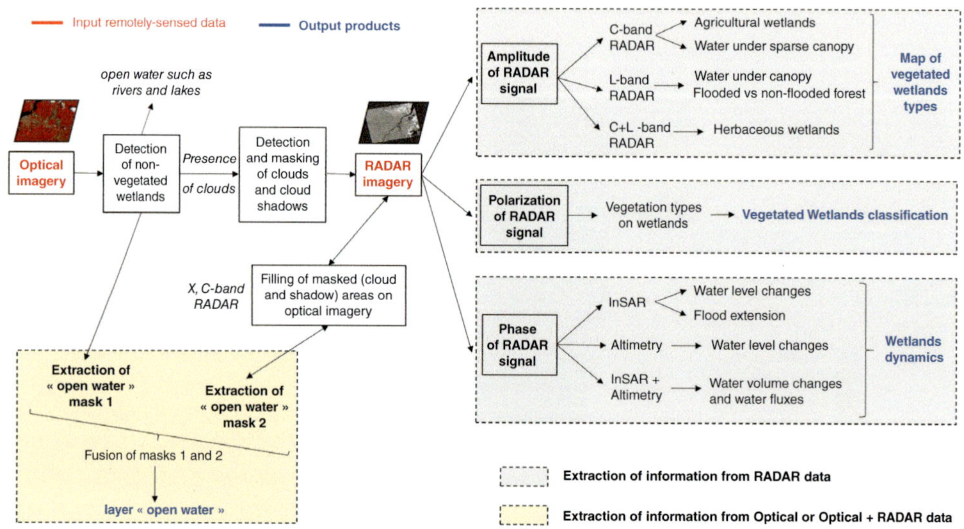

Fig. 2.1.2. Framework of combining optical and SAR remotely sensed data for characterizing and mapping wetlands and accumulations of water. (From: Catry *et al*., 2018a.)

For instance, *Anopheles* mosquitoes depend, to some degree, on the presence of forested areas and strongly depend on the presence of water for their survival and propagation. However, deforested areas provide favorable conditions for malaria vector breeding and feeding, and forest and secondary forest provide resting sites for adult mosquitoes after feeding (Yasuoka and Levins, 2007; Vittor *et al.*, 2009; Hahn *et al.*, 2014; MacDonald and Mordecai, 2019).

Challenges and questions

The tasks of identifying and quantifying environmental determinants involved in the transmission of mosquito-borne diseases are the main challenges and opportunities for the use of EO data in public health. In addition, researchers need to gain a better understanding of how these determinants relate to socio-economic, socio-demographic, and human behavioral factors. Depending on scope and purpose, assessments of mosquito-borne disease risks require EO data at various levels of detail, ranging from very high to moderate spatial resolution, and at various temporal scales, involving seasonal to daily data acquisition. Many environmental variables can be derived from EO data streams, including temperature, humidity, wind and wind speed, as well as land use and land cover information. For detailed geospatial mosquito habitat assessment, several thematic data sources need to be collated. These can be used to gauge the impact of actual weather conditions, to map land use and land cover, and to relate the information to settlement locations, exposure, built-up area configurations, and behavioral patterns of the local population. On one hand, previous studies have noted insufficient EO data for the composition of coherent time series and the lack of accessible very-high-resolution data or SAR data (Herbreteau *et al.*, 2007; Machault *et al.*, 2011). The high cost for very high spatial resolution satellite data for producing adequate spatial coverage is a barrier for the R&D use of such data and its application in public health programs. On the other hand, researchers and practitioners are faced with mounting data assimilation and processing demands and a dearth of available processing capabilities.

There are several questions and critical issues that need to be answered and resolved, including:

- What EO data sets are most suitable, accessible, and practical for producing risk maps of mosquito-borne disease transmission, i.e. for identifying where mosquito-borne disease transmission can occur?

- What EO data sets are most suitable, accessible, and practical for determining seasonal or weekly changes in risk associated with changes in mosquito density and infection, i.e. for forecasting risk on a weekly to monthly basis associated with rates of mosquito reproduction and mortality and development rates of pathogens in mosquitoes?
- What are the main constraints in terms of obtaining, maintaining, and delivering EO-derived products and services to researchers, public health policy makers, and practitioners involved in mosquito-borne disease control programs?

Responses and options

Below are the comments and suggestions of the experts consulted about critical issues and EO data requirements in the study and analysis of mosquito-borne diseases:

- Objectives requiring timely geospatial information on mosquito habitats can be achieved with EO-based land cover and land use mapping, with a focus on urban and agricultural areas.
- Objectives based on information on mosquito abundance require timely EO-derived information on temperature, humidity, precipitation, and suitable environment, and require mosquito distribution maps at various spatial resolutions.
- In some instances, a combination of optical, thermal, and SAR data may be needed.
- The spatial and temporal resolution of EO data required to develop risk maps for public health needs to match weather and environmental determinants that drive, in part, the transmission of mosquito-borne diseases. There is a need to characterize and identify, at a local scale, areas of high spatial and temporal mosquito density; medium to high spatial resolution is required for identifying mosquito habitat areas.
- There is a need for multi-temporal EO data acquisitions, selection of complementary data sets, skillful application of image processing techniques, and allocation of sufficient financial resources to accomplish the above.

Modeling environment–human–vector interaction hazard using EO data and land cover maps in a local, cross-border setting between French Guiana and Brazil

The prevention and control of mosquito-borne diseases are challenging public health issues. Disease transmission is a multi-scale process, strongly controlled by weather and environmental factors. Remote sensing data analyses are suitable for characterizing spatial and temporal dynamics of such diseases. Yet, despite the growing number of EO data sources and various technical capacities currently available, the selection of suitable EO data for the production of hazard maps and exposure risk maps remains a challenging task. The crucial issue is the selection of adequate EO-derived geospatial time series that fit the temporal and spatial dynamics of the studied disease.

We present here as a case study the research of Li *et al.* (2016), in which the role of land cover classes involved in the life cycle of the malaria vector (*Anopheles darlingi*) in the Amazon region was investigated. SPOT 5 (Satellite pour l'Observation de la Terre 5), optical satellite imagery taken in 2012 at 10 m resolution was used to produce a land cover map from which landscape indicators were derived, including forest fragmentation and density of boundaries between forested and non-forested areas (Fig. 2.1.3).

The study relied on partial knowledge-based modeling of malaria transmission risk for a 500 km^2 area in the Amazon region between French Guiana and Brazil, using a landscape-based approach and review of pertinent literature. A landscape model was obtained by generating land use and land cover (LULC) maps of the area, followed by computing and combining landscape metrics to build a set of normalized landscape-based hazard indices. The quantitative landscape characterization involved defining a spatial window for the metrics computation. The dimension of this window corresponds to a zone where the landscape characteristics are

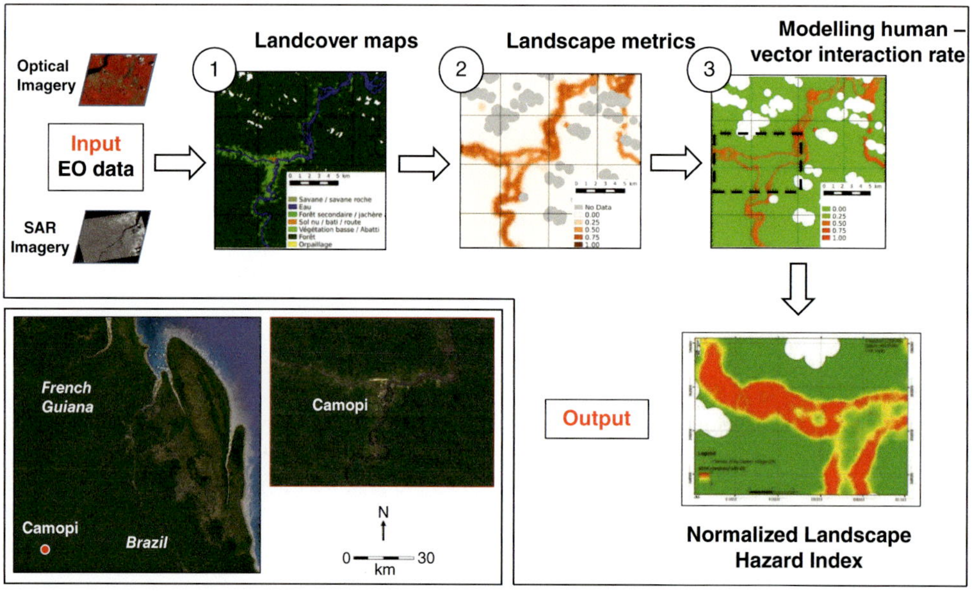

Fig. 2.1.3. Flow chart outlining vector-human interaction hazard mapping in the study of Malaria, with land cover classification derived from optical and SAR EO data for the Camopi area in the border region between French Guiana and Brazil in South America. (From: Li *et al.*, 2016).

most likely to influence the chance of encounter between *Anopheles* mosquitoes and human beings. A Normalized Landscape-based Hazard Index (NLHI) was selected in conjunction with the knowledge-based model and connection with incidence of malaria caused by *Plasmodium falciparum* (Li *et al.*, 2016).

Analysis results revealed that hazard-free areas (green color on index map in Fig. 2.1.3) around the village of Camopi consist of dense forest areas that are not affected by deforestation and areas where the anthropogenic pressure is high, for example at the confluence of two rivers. Conversely, high hazard areas (yellow and red colors) correspond to the areas where there is a high density of forest edge and where the percentage of forest is higher than in the zones with the highest anthropogenic pressure. Li *et al.* (2016) validated this approach with actual malaria incidence from the cross-border region between French Guiana and Brazil. This study confirms that EO data can be an efficient tool for identifying environmental features related to malaria transmission and that an NLHI of malaria transmission can be developed using satellite imagery.

However, the presence of clouds and cloud shadows in many tropical environments results in missing data on optical images. Likewise, many wetland areas that are obscured from view by vegetation canopies – and hence are not observable by optical remote sensing – can conceivably contain breeding sites for malaria vectors. Alternatively, SAR can be used or combined with optical imagery for extracting environmental information related to vector habitats, as SAR has proven itself capable of penetrating clouds and detecting water bodies reliably. Further research should consider the temporal aspects of deforestation by producing a time series of land cover maps and then studying the evolution of the NLHI associated with malaria in the Amazon region. Institutions in Brazil, such as the National Institute for Space Research (INPE), already produce such deforestation maps derived from satellite imagery under Project PRODES.[2] Li *et al.* (2017) demonstrated that the NLHI calculation can be scaled up from a local scale to a regional scale. The NLHI calculations for the Amazon region currently involve biomass map products[3] of Landsat-based deforestation time series over the Brazilian territory.

Expected outcomes and impacts

This study establishes a malaria hazard index that is driven by spatial knowledge and landscape information using EO data as an important input. The index can be produced on a regular basis in support of malaria prediction, surveillance, and control. The index is calculated using LULC maps as input in the geospatial model; the model output maps serve actors of disease surveillance and vector control (Fig. 2.1.3). These maps identify areas where interactions between malaria vectors and human populations are likely to occur, based on the spatial configuration of landscape features. In essence, the maps provide information on locations where people are more likely to be exposed to mosquitoes and infected by malaria pathogens. This is a key element to take into account when defining and optimizing vector control strategies for public health responses.

The example presented here shows an application of EO data to health issues at a local scale. This approach was subsequently applied at a regional scale (Li *et al.*, 2017) and is currently extended to include the entire Amazon region. Since this region covers more than 6.5 million km^2 and spans nine countries (Bolivia, Brazil, Colombia, Ecuador, France/French Guiana, Guyana, Peru, Suriname, and Venezuela), working at this scale requires the use of EO data in order to produce the required geospatial information and address public health issues. It is extremely difficult to deal with the health problems of different countries when data sources are heterogeneous in terms of content and quality. The example of the cross-border area between French Guiana and Brazil demonstrates that EO data analysis is an effective way to produce homogeneous and standardized information that overcomes this problem. Updates of these maps are possible using multi-temporal EO data – weekly, monthly, or seasonal EO-based updates can be provided depending on the satellite and sensor system selected. In fact, current SAR and optical imagery from the Sentinel Constellations and the European Space Agency's (ESA) Copernicus Programme provide weekly data free of charge at a spatial resolution that is adequate for such large-scale cross-border applications of EO for health issues.

The end users for such maps are actors in the public health domain representing local, regional, and national institutions. More specifically, the primary users of these maps are concerned with the elaboration of vector control strategies and activities in the field. EO data can potentially bridge part of the information gap that confronts health surveillance communities. Yet, going beyond the scope and content of the case studies presented here, the needs of public health actors in terms of various geospatial data and products are not always satisfied for two reasons. First, satellite sensors are not primarily designed for health applications, often rendering spatial, temporal, or spectral data properties inadequate for addressing public health issues. Second, the methodologies for the production of hazard and risk maps developed by researchers of the EO community may not always be suitable or adequate in a public health context due to the complexity of the methodologies, the cost of high-resolution data, and the lack of computing resources.

Technical considerations and perspectives for producing risk maps

The production of LULC maps and hazard maps like those shown in Fig. 2.1.1 and Fig. 2.1.2 requires optical and SAR images at various spatial and temporal resolutions. In this case, environmental variables are extracted from three different sources.

High-resolution EO products with high temporal resolution, including Sentinel-1 and ALOS SAR data and Sentinel-2 optical data, are the primary products needed for the generation of these maps. Data access is free and data acquisition can occur worldwide every 5–12 days. The high-resolution products can be complemented with very-high-resolution imagery, albeit less frequently. For instance, optical sensors of the French Pléiades satellite constellation can acquire images at 50 cm resolution. Although extremely useful for detailed studies of mosquito-borne diseases within urban environments, in an operational context, the cost and volume of such data could prove prohibitive. Commercially available SAR data are also very expensive. Lower resolution images from the advanced very high resolution radiometer (AVHRR), Moderate Resolution Imaging Spectroradiometer (MODIS) or Visible/Infrared Imager Radiometer Suite (VIIRS) sensors are a suitable source for identifying microclimatic indicators related to variables like surface temperature, surface moisture, near-surface air temperature, and water stress; these data are acquired daily and can be accessed without charge.

Frequent updates and cloud presence require the use of a multi-temporal series of optical EO data and the combination of optical and SAR data. This necessitates considerable data storage resources for regular production of land cover and risk maps, and for their use in an operational context. The addition of sensors recently launched (such as the RADARSAT Constellation Mission), or future launches such as the Surface Water Ocean Topography (SWOT) satellite planned for 2022 and the BIOMASS for 2023, will increase the volume of EO data utilization and attendant data storage issues. Future developments in EO big data storage and sharing will also have to take these aspects into account and possibly rely on cloud computing for data storage, processing, and analysis. Following the model currently proposed by "Google Earth Engine," large volumes of data could be remotely processed and analyzed without downloading the data.

Using EO big data implies the development of adapted computing methods such as artificial intelligence and machine learning algorithms. Together with storage capacities, computing resources will have to be customized for such applications. Automated and generic methods are preferred as they would facilitate the production of EO-based products like land cover maps anywhere in the world. Likewise, the analysis of images and the production of the risk maps require image processing software and a GIS capability. Expertise in EO image analysis, geo-informatics, and mapping is essential for the development of risk maps. Availability of freeware and EO and GIS "toolboxes," open access to EO data, as well as training programs strongly encourage the use of EO products by non-specialists, including those in the public health sector. Note that the Copernicus RUS (Research and User Support) service portal, managed by the ESA, offers assistance to users. The portal promotes the uptake of Copernicus data and helps the scaling up of R&D activities with its data. They also offer free access to computing resources, storage, and freeware for processing data and developing technical solutions customized to users' needs, they provide a dedicated helpdesk for assistance, and they organize regular training sessions.[4]

Many new sensors are to be launched in the next few years, offering new possibilities in terms of spatial and temporal resolutions, and technical capabilities. Together with the currently orbiting high- and moderate-resolution sensors, the RADARSAT Constellation Mission and the SWOT and BIOMASS missions, among others, will provide new EO data sources to produce more accurate land cover maps, time series, and quality information for vector control and surveillance. While EO products and methodologies will initially have to be custom designed to better fit public health needs, proven methodologies need to be automated in the future and be robust and user-friendly enough to be implemented by non-specialists. To do so, the remote sensing, entomology, epidemiology, and public health communities have to interact more efficiently. They need to form a community of practice, integrating data from a wide variety of sources at various scales and qualities to help mitigate public health issues such as mosquito-borne diseases.

Risk mapping of entomological Rift Valley fever in Senegal at high spatio-temporal resolution using remote sensing

The emergence and re-emergence of infectious diseases with high epidemic potential, such as Rift Valley fever (RVF), have caused public health actors to adapt their management strategies concerning human and veterinary health. RVF is transmitted by mosquitoes and is naturally maintained by wildlife reservoir hosts. In outbreak situations, transmission cycles among wildlife spill over into livestock. Humans can acquire infections from mosquitoes but also from infected livestock. This adaptation requires the development of new means of risk prediction. In this context, the study of vector-borne infectious diseases requires the knowledge of factors conducive to the emergence and spread of those diseases.

The French space agency Centre national d'études spatiales (CNES) and its partners have applied the conceptual approach of tele-epidemiology to RVF (Fig. 2.1.4). Factors determining the occurrence and spread of pathogens can be environmental, climatic, demographic, socio-economic, and/or behavioral. Some can be identified by EO data analysis, which requires the development of effective methods to use remote sensing for risk factor characterization, mapping, and monitoring. This methodological approach has been successfully applied to RVF in the Ferlo region of Senegal, leading toward the development of a dynamic mapping procedure of Zones Potentially Occupied by Mosquitoes (ZPOMs) (Lacaux *et al.*, 2007). RVF is a viral disease that occurs largely in Africa, causing very serious economic losses in livestock.

The RVF project presented here depends on the cooperation of French and Senegalese institutions, including the Centre de Suivi Ecologique, the Dakar Pasteur Institute, the Direction of Veterinarian Services, Météo-France, and CNES (Lafaye *et al.*, 2013). The project has developed a new decision support tool utilizing SPOT-5 satellite imagery with the objective to improve animal health management and support local users in the public health sector. Funding support was provided by the French Ministry of Ecology.

In the Ferlo region of Senegal, the abundance of the main RVF vectors (*Aedes vexans* and *Culex poicilipes*) is directly linked to the occurrence and extent of surface water ponding, which is closely related to the spatio-temporal variability of rainfall events

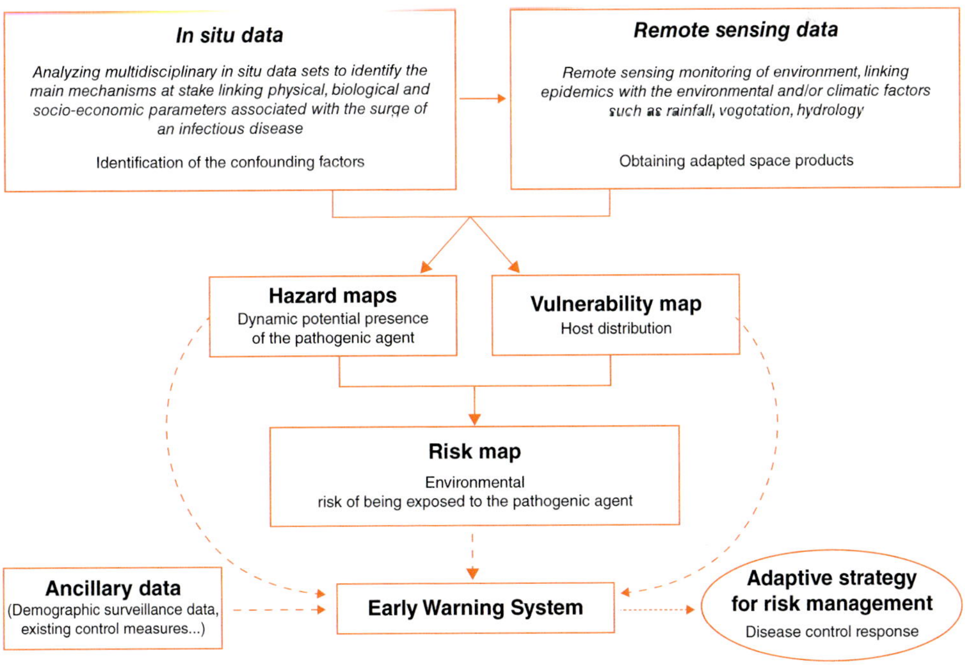

Fig. 2.1.4. The conceptual approach of tele-epidemiology for vector-borne diseases.

(Guilloteau *et al.*, 2014). Hence, rainfall distribution and its spatial heterogeneity is a key parameter for the emergence of the main RVF vectors. The goal of the project was to use GIS tools and EO data to detect ponds as potential breeding sites and evaluate the risk of exposure for cattle to vector bites. A risk model for the emergence of mosquitoes has been developed and validated using field entomological surveillance (Bicout *et al.*, 2003, 2015; Porphyre *et al.* 2005).

Three steps have been necessary to achieve the goal. As a first step, a procedure and index were established for detecting and mapping small and temporary ponds with high-resolution SPOT-5 imagery. Repeat satellite data acquisitions provided synoptic views concerning the dynamics of the approximately 1300 ponds as potential vector breeding sites in the Barkédji area. A Normalized Difference Pond Index (NDPI) was obtained by combining data of the green and short-wave infrared (SWIR) bands.

The second step involved modeling ZPOMs by linking rainfall variability, pond dynamics, and density of aggressive vectors. Spot-5 images and meteorological information from *in situ* data collection or data from five satellite-based rainfall products – Tropical Rainfall Measuring Mission (TRMM), Global Satellite Mapping of Precipitation (GSMaP), African rainfall estimate (RFE), Climate Prediction Center morphing method (CMORPH), and Precipitation Estimation from Remotely Sensed Information using Artificial Neural Networks (PERSIANN) – were used to fit a model with hydrological and entomological components. The modeling results consisted of dynamic maps that were generated on a daily basis at a spatial resolution of 10 m to predict the entomological risk for RVF in the Ferlo region of Senegal (Fig. 2.1.5).

The third step consisted of overlaying vector hazard information in the form of the dynamic ZPOMs and host vulnerability information in the form of the location of beef feedlot cattle grazing area to evaluate the environmental risk of cattle exposure to vector bites. Integrating the dynamic model on mosquito proliferation and the position of actual livestock grazing areas into a GIS allowed the Directorate of Veterinary Services of Senegal to issue, on a trial basis, weekly risk zone forecasting bulletins valid for the subsequent 10 days.

Expected outcomes and impacts

The maps generated by this project indicate and outline the RVF risk areas associated with surface water ponding, mosquito breeding, and cattle grazing for a test area in Senegal. EO satellite data offered synoptic views and repeated measurements concerning the location and extent of more than 1300 ponds. The scope and frequency of this undertaking would not have been feasible by means of *in situ* data collection.

The end user of the RVF project products is the Directorate of Veterinary Services of Senegal, who can integrate this information into its adaptation strategy of animal health management. This strategy could include the following recommendations to effectively mitigate the exposure of cattle to RVF, and thus to minimize infection risk for humans:

- (Re-)locate livestock grazing areas away from risk zones, with warning signs in local languages posted near the ponds to inform breeders to keep their animals at least 500 m away from the ponds.
- Issue regular bulletins so the Pasteur Institute of Dakar can organize efficient larval and vector control actions.
- Issue regular bulletins so the Directorate of Veterinary Services of Senegal can organize and optimize vaccination campaigns in the riskiest zones.
- Establish a joint communication strategy by integrating information of the forecasted risk bulletins into the National Information System of Surveillance of Epidemics used by the Ministry of Livestock in Senegal and the Headquarters of the Directorate of Veterinary Services of Senegal and its local representatives in rural districts.
- Plan to broadcast RVF-related messages in local languages through local radio stations.

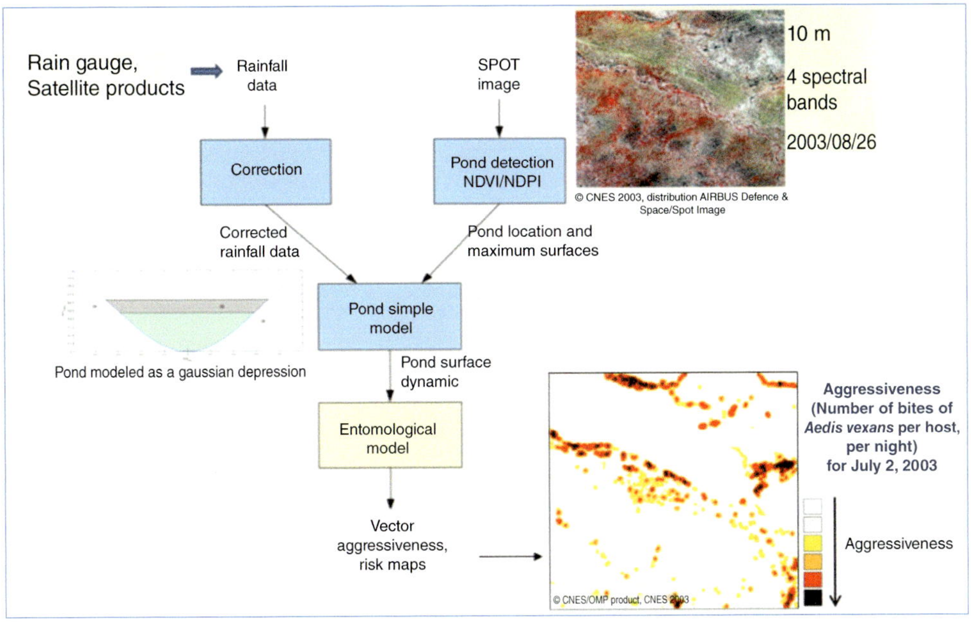

Fig. 2.1.5. Flow chart outlining the RVF entomological risk modeling approach.

Technical considerations and perspectives for producing risk maps

The RVF maps have been built based on a value chain proposition that clearly identifies satellite data sources and data provider, the service provider, and the end user (Table 2.1.1). EO data analysis and the production of the risk map require image processing GIS software packages. Expertise in EO image analysis, geo-informatics, and mapping is essential for the production of risk maps.

In the absence of SPOT-5, which ceased operation, the opportunity exists to access Sentinel-2 satellite data for mapping rain-fed ponds in the manner proposed by the RVF tool. The constellation of the Sentinel-2A and Sentinel-2B satellites could deliver images with adequate spectral, spatial, and temporal resolution required to produce the risk maps at the scale that meets the needs of the user. Future development should consider the implementation of this tool through an open-source software. The following table lists examples of EO-derived products that are potentially useful as geospatial reference or background formation for public health-related studies and applications. While these products have not been devised initially with public health applications in mind, they could provide important resources and insights for the understanding of mosquito-borne disease dynamics (Table 2.1.2).

Notes

[1] http://www.who.int/campaigns/world-health-day/2014/global-brief/en/, see also World Meteorological Organization (WMO) 2014.

[2] http://www.obt.inpe.br/OBT/assuntos/programas/amazonia/prodes (accessed 31 December 2021).

[3] http://mapbiomas.org (accessed 31 December 2021).

[4] https://rus-copernicus.eu/portal/ (accessed 31 December 2021).

Table 2.1.1. The value chain of the RVF project.

Satellites →	Data provider →	Service provider →	End user →	Benefit
		Centre de Suivi Ecologique de Dakar	Directorate of Veterinary services of Senegal	*Better management of animal health*
SPOT-5 ===➤	Optical image By Airbus Defense and Space	Small and temporary pond mapping at 10 m resolution	End user adapts and optimizes their strategy of animal health management	
TRMM ===➤ GPM-core GCOM-W-AMSR2 DSMP-SSMI NOAA-AQUA NOAA-AMSU METOP-AMSU GOES-8 GOES-10 Meteosat-6 Meteosat-7	Satellite rainfall estimates TMPA (TRMM Multi-satellite Precipitation Analysis) by NASA/JAXA GSMap (Global Satellite Mapping of Precipitation) products by JAXA-CREST RFE (African Rainfall Estimation) by NOAA-CPC PERSIANN (Precipitation Estimation from Remotely Sensed Information Using Artificial Neural Networks) by the CHRS, University of California CMORPH product from the DMSP, NOAA, Aqua, and TRMM satellites by NOAA-CPC **Ground data** Entomological data by the Dakar Pasteur Institute	Dynamic high-resolution maps (10 m spatial resolution, daily temporal resolution) predicting the entomological risk for Rift Valley fever (presence of mosquitoes) Forecasting bulletins of risk zones for cattle exposed to mosquito bites		

AMSU, Advanced Microwave Sounding Unit; AQUA, Aqua Earth-observing satellite mission; CHRS, Center for Hydrometeorology and Remote Sensing (University of California); NOAA CMORPH, Climate Prediction Center morphing method; CPC, Climate Prediction Center; DMSP, NOAA Defense meteorological satellite program; GCOM-W-AMSR2, Global Change Observation Mission – Water "Shizuku" – Advanced Microwave Scanning Radiometer 2; GOES, Geostationary Satellite Server; GPM, global precipitation measurement mission; JAXA, Japan Aerospace Exploration Agency; Metop, meteorological operational satellite; NOAA, National Oceanic and Atmospheric Administration; SPOT 5, Satellite pour l'Observation de la Terre 5; SSMI, special sensor microwave imager; TRMM, tropical rainfall measuring mission.

Table 2.1.2. Examples of EO-derived products that are potentially useful as geospatial reference or background formation for public health-related studies and applications.

Product type	Application in public health
Global land cover maps (e.g., MERIS GlobCover, PALSAR forest vs. non-forest maps, SAR global wetland maps)	For coarse identification of environmental features and habitat suitability to vectors for targeted studies
Vegetation indices (NDVI or EVI from MODIS or AVHRR)	For showing the evolution of vegetation cover (deforestation) and its implications on the distribution of vectors
Soil moisture (SMOS)	For mapping potential breeding sites for some mosquito species

Continued

Table 2.1.2. Continued.

Product type	Application in public health
Continental water quality maps from MODIS	For assessing the suitability of water and wetlands to the development of mosquito larvae (potential breeding sites)
DEMs from SRTM or TandDEM-X	For assessing the role of topography on water circulation and breeding site distributions
Time series of EO products	For assessing the dynamics of the relationships between environmental features and disease transmission
Meteorological sensors	For assessing the role of climate variables on disease transmission
Climate models	For providing scenarios and predicting disease distributions worldwide

AVHRR, advanced very high-resolution radiometer; DEM, Digital Elevation Model; EO, Earth Observation; EVI, Enhanced Vegetation Index; MERIS GlobCover, Medium Resolution Imaging Spectrometer, Global land cover; MODIS, Moderate Resolution Imaging Spectroradiometer; NDVI, Normalized Difference Vegetation Index; PALSAR, Phased Array L-band Synthetic Aperture Radar; SAR, synthetic aperture radar; SMOS, Soil Moisture Ocean Salinity; SRTM, Shuttle Radar Topography Mission; TandDEM-X, TerraSAR-X add-on for Digital Elevation Measurement.

References

Baeza, A., Bouma, M.J., Dhiman, R.C., Baskerville, E.B., Ceccato, P. *et al.* (2013) Long-lasting transition toward sustainable elimination of desert malaria under irrigation development. *Proceedings of the National Academy of Science* 110, 15157–15162. DOI: https://doi.org/10.1073/pnas.1305728110.

Bejon, P., Williams, T.N., Liljander, A., Noor, A.M., Wambua, J. *et al.* (2010) Stable and unstable malaria hotspots in longitudinal cohort studies in Kenya. *PLoS Medicine* 7. DOI: https://doi.org/10.1371/journal.pmed.1000304.

Bicout, D.J., Porphyre, T., Ndione, J.-A., and Sabatier, P. (2003) Modelling abundance of *Aedes* and *Culex* spp. in rain fed ponds in Barkedji, Senegal. Proceedings of the 10th Symposium of the International Society for Veterinary Epidemiology and Economics/*ISVEE 10*, Vina del Mar, Chile, 84. http://www.sciquest.org.nz/node/63596 (accessed 14 June 2021).

Bicout, D.J., Vautrin, M., Vignolles, C., and Sabatier, P. (2015) Modeling the dynamics of mosquito breeding sites versus rainfall in Barkedji area, Senegal. *Ecological Modelling* 317, 41–49. DOI: https://doi.org/10.1016/j.ecolmodel.2015.08.027.

Catry, T., Li, Z., Roux, E., Herbreteau, V., Dessay, N. *et al.* (2016) Fusion of SAR and optical imagery for studying the eco-epidemiology of vector-borne diseases in tropical countries. *Proceedings of the ESA Living Planet Symposium, Prague, Czech Republic*. Available at: https://hal.archives-ouvertes.fr/hal-01490830 (accessed 14 June 2021).

Catry, T., Li, Z., Roux, E., Herbreteau, V., Gurgel, H. *et al.* (2018a) Wetlands and malaria in the Amazon: Guidelines for the use of synthetic aperture radar remote sensing. *International Journal of Environmental Research and Public Health* 15, 468. DOI: https://doi.org/10.3390/ijerph15030468.

Catry, T., Pottier, A., and Marti, R. (2018b) Apports de la combinaison d'images satellites optique et RADAR dans l'étude des maladies à transmission vectorielle: cas du paludisme à la frontière Guyane française – Brésil. *Confins* 37. DOI: https://doi.org/10.4000/confins.15027.

Ceccato, P., Connor, S.J., Jeanne, I., and Thomson, M.C. (2005) Application of geographical information systems and remote sensing technologies for assessing and monitoring malaria risk. *Parassitologia* 47, 81–96. Available at: https://www.researchgate.net/publication/7700360_Application_of_Geographical_Information_Systems_and_Remote_Sensing_technologies_for_assessing_and_monitoring_malaria_risk (accessed 5 August 2021).

Ewing, D.A., Purse, B.V., Cobbold, C.A., and White, S.M. (2021) A novel approach for predicting risk of vector-borne disease establishment in marginal temperate environments under climate change:

West Nile virus in the UK. *Journal of the Royal Society Interface* 18. DOI: https://doi.org/10.1098/rsif.2021.0049.

Ferraguti, M., Martínez-de la Puente, J., and Figuerola, J. (2021) Ecological effects on the dynamics of West Nile virus and avian Plasmodium: The importance of mosquito communities and landscape. *Viruses* 13, 1208. DOI: https://doi.org/10.3390/v13071208.

Franke, J., Gebreslasie, M., Bauwens, I., Deleu, J., and Siegert, F. (2015) Earth observation in support of malaria control and epidemiology: MALAREO monitoring approaches. *Geospatial Health* 10, 117–131. DOI: https://doi.org/10.4081/gh.2015.335.

Guilloteau, C., Gosset, M., Cecile Vignolles, C., Alcoba, M., Tourrem Y.M. *et al.* (2014) Impacts of satellite-based rainfall products on predicting spatial patterns of Rift Valley fever vectors. *Journal of Hydrometeorology* 15, 1624–1635. DOI: https://doi.org/10.1175/JHM-D-13-0134.1.

Hahn, M.B., Gangnon, R.E., Barcellos, C., Asner, G.P., and Patz, J.A. (2014) Influence of deforestation, logging, and fire on malaria in the Brazilian Amazon. *PLoS ONE* 9. DOI: https://doi.org/10.1371/journal.pone.0085725.

Hay, S.I., Snow, R.W., and Rogers, D.J. (1998a) From predicting mosquito habitat to malaria seasons using remotely sensed data: practice, problems and perspectives. *Parasitology Today* 14, 306–313. DOI: https://doi.org/10.1016/s0169-4758(98)01285-x.

Hay, S.I., Snow, R.W., and Rogers, D.J. (1998b) Predicting malaria seasons in Kenya using multitemporal meteorological satellite sensor data. *Transactions of the Royal Society of Tropical Medicine and Hygiene* 92, 12–20. DOI: https://doi.org/10.1016/s0035-9203(98)90936-1.

Hay, S.I. and Lennon, J.J. (1999) Deriving meteorological variables across Africa for the study and control of vector-borne disease: A comparison of remote sensing and spatial interpolation of climate. *Tropical Medicine and International Health* 4, 58–71. DOI: https://doi.org/10.1046/j.1365-3156.1999.00355.x.

Herbreteau, V., Salem, G., Souris, M., Hugot, J.P., and Gonzalez, J.P. (2007) Thirty years of use and improvement of remote sensing applied to epidemiology: From early promises to lasting frustration. *Health & Place* 13, 400–403. DOI: https://doi.org/10.1016/j.healthplace.2006.03.003.

Kalluri, S., Gilruth, P., Rogers, D., and Szczur, M. (2007) Surveillance of arthropod vector-borne infectious diseases using remote sensing techniques: A review. *PLoS Pathogens* 3, 1361–1371. DOI: https://doi.org/10.1371/journal.ppat.0030116.

Kotchi, S.O., Bouchard, C., Ludwig, A., Rees, E.E., and Brazeau, S. (2019) Using Earth observation images to inform risk assessment and mapping of climate change-related infectious diseases. *Canada Communicable Disease Report* 45, 133–142. DOI: https://doi.org/10.14745/ccdr.v45i05a04.

Lacaux, J.-P., Tourre, Y.M., Vignolles, C., Ndione, J.A., and Lafaye, M. (2007) Classification of ponds from high-spatial-resolution remote sensing: Application to Rift Valley fever epidemics in Senegal. *Remote Sensing of Environment* 106, 66–74. DOI: https://doi.org/10.1016/j.rse.2006.07.012.

Lafaye, M., Sall, B., Ndiaye, Y., Vignolles, C., Tourre, Y.M. *et al.* (2013) Rift Valley fever dynamics in Senegal: A project for pro-active adaptation and improvement of livestock raising management. *Geospatial Health* 8, 279–288. DOI: https://doi.org/10.4081/gh.2013.73.

Li, Z., Roux, E., Dessay, N., Girod, R., Stefani, A. *et al.* (2016) Mapping a knowledge-based malaria hazard index related to landscape using remote sensing: Application to the cross-border area between French Guiana and Brazil. *Remote Sensing* 8, 319. DOI: https://doi.org/10.3390/rs8040319.

Li, Z., Catry, T., Dessay, N., da Costa Gurgel, H., Aparecido de Almeida, C. *et al.* (2017) Regionalization of a landscape-based hazard index of malaria transmission: An example of the State of Amapá, Brazil. *Data* 2, 37. DOI: https://doi.org/10.3390/data2040037.

Lucas, R., Blonda, P., Bunting, P., Jones, G., Inglada, J. *et al.* (2015) The Earth Observation Data for Habitat Monitoring (EODHaM) system. *International Journal of Applied Earth Observation and Geoinformation* 37, 17–28. DOI: https://doi.org/10.1016/j.jag.2014.10.011.

MacDonald, A.J and Mordecai E.A. (2019) Amazon deforestation drives malaria transmission, and malaria burden reduces forest clearing. *Proceedings of the National Academy of Sciences* 116, 22212–22218. DOI: https://doi.org/10.1073/pnas.1905315116.

Machault, V., Vignolles, C., Borchi, F., Vounatsou, P., and Briolant, S. (2011) The use of remotely sensed environmental data in the study of malaria. *Geospatial Health* 5, 151–168. DOI: https://doi.org/10.4081/gh.2011.167.

Machault, V., Yébakima A., Etienne M., Vignolles C., Palany P. *et al.* (2014) Mapping entomological dengue risk levels in Martinique using high-resolution remote sensing environmental data. *ISPRS: International Journal of Geo-Information* 3, 1352–1371. DOI: https://doi.org/10.3390/ijgi3041352.

Moua, Y., Kotchi, S.O., Ludwig, A., and Brazeau, S. (2021) Mapping the habitat suitability of West Nile virus vectors in southern Quebec and eastern Ontario, Canada, with species distribution modeling and satellite earth observation data. *Remote Sensing* 13, 1637. DOI: https://doi.org/10.3390/rs13091637.

Porphyre, T., Bicout, D.J., and Sabatier, P. (2005) Modelling the abundance of mosquito vectors versus flooding dynamics. *Ecological Modelling* 183, 173–181. DOI: https://doi.org/10.1016/j.ecolmodel.2004.06.044.

Rogers, D.L., Randolph, S.E., Snow, R.W., and Hay, S.I. (2002) Satellite imagery in the study and forecast of malaria. *Nature* 415, 710–715. DOI: https://doi.org/10.1038/415710a.

Tatem, A.J. and Hay, S.I. (2004) Measuring urbanization pattern and extent for malaria research: A review of remote sensing approaches. *Journal of Urban Health* 81, 363–376. DOI: https://doi.org/10.1093/jurban/jth124.

Vittor, A.Y., Pan, W., Gilman, R.H., Tielsch, J., Glass, G. *et al.* (2009) Linking deforestation to malaria in the Amazon: Characterization of the breeding habitat of the principal malaria vector, *Anopheles darlingi*. *The American Journal of Tropical Medicine and Hygiene* 81, 5–12. Available at: https://www.researchgate.net/publication/26320834_Linking_Deforestation_to_Malaria_in_the_Amazon_Characterization_of_the_Breeding_Habitat_of_the_Principal_Malaria_Vector_Anopheles_darlingi (accessed 5 August 2021).

World Health Organization (WHO) (2012) World malaria report 2012. Available at: https://www.who.int/malaria/publications/world_malaria_report_2012/wmr2012_full_report.pdf (accessed 14 June 2021).

World Health Organization (WHO) (2015) Global technical strategy for malaria 2016–2030. Available at: https://www.who.int/docs/default-source/documents/global-technical-strategy-for-malaria-2016-2030.pdf?sfvrsn=c82afcc_0 (accessed 14 June 2021).

Yasuoka, J. and Levins, R. (2007) Impact of deforestation and agricultural development on anopheline ecology and malaria epidemiology. *The American Journal of Tropical Medicine and Hygiene* 76, 450–460. DOI: https://doi.org/10.4269/ajtmh.2007.76.450.

2.2 Tick-borne Diseases

Nicholas H. Ogden[1], Serge Olivier Kotchi[1], Stéphanie Brazeau[1], Catherine Bouchard[1], Joanne C. White[2a], Michael A. Wulder[2a], Andrew Davidson[3], André Beaudoin[2b], and Dirk Werle[4]

[1]Public Health Risk Sciences Division, National Microbiology Laboratory, Public Health Agency of Canada, Saint-Hyacinthe, Canada; [2a]Canadian Forest Service, Natural Resources Canada, Victoria, Canada; [2b]Canadian Forest Service, Natural Resources Canada, Québec, Canada; [3]Science and Technology Branch, Agriculture and Agri-Food Canada, Ottawa, Canada; [4]Ærde Environmental Research, Halifax, Canada

Context, state of knowledge, challenges, and responses

Lyme disease is the most commonly reported vector-borne disease in the northern temperate climatic zone and occurs in North America, Europe, and Asia. Lyme disease is an emerging infectious disease in Canada due to the northward spread of the geographic range of the tick vector *Ixodes scapularis* from the USA. Lyme disease is an environmental health hazard that occurs where the environment, in terms of climate and woodland habitat, is suitable for the ticks and the natural wildlife hosts of both the ticks and the causal bacterium *Borrelia burgdorferi*. Risk factors for Lyme disease are mostly behaviors and outdoor activities that increase the risk of exposure to tick bites, such as people coming into the woodland environments where the ticks are found (Aenishaenslin *et al.*, 2017; Bouchard *et al.*, 2018). There is evidence that climate change is driving the spread of these ticks in Canada, and the geographic range of endemic Lyme disease risk in Canada is expected to increase as the climate warms (Ogden *et al.*, 2008, 2014; Leighton *et al.*, 2012; Clow *et al.*, 2017). Several landscape, climatic, and environmental (habitat) conditions have to converge for the co-existence of hosts, tick vectors, and transmission cycles of *B. burgdorferi* for significant Lyme disease risk to occur. Consequently, risk maps based on geospatial information on environmental conditions (particularly climate and habitat) and vector dispersion are being developed to assist public health in targeting surveillance efforts and developing prevention programs (Kotchi *et al.*, 2021).

Examples of recent research

Studies in Canada and Europe have focused on mapping where tick vectors and risk to the public occur. Ticks have long life cycles and, once they have become established, inter-annual variations in risk are small relative to those that can occur for mosquito-borne diseases. Consequently, understanding environmental drivers of Lyme disease risk focuses more on predicting *where* risk is likely to be now or in the future, rather than on early warning forecasting to identify *when* outbreaks may occur, which is often a public health objective in the context of mosquito-borne diseases (Ogden and Lindsay, 2016). Bouchard *et al.* (2015) reviewed Canadian studies that connect the spatio-temporal pattern of Lyme disease emergence and expansion to a number of key environmental drivers. These include climate, habitat, and wild animal host community conditions that determine environmental suitability for vectors and pathogen transmission cycles, and patterns of dispersion of ticks and pathogens by migratory birds and other animal hosts.

Both ground-level and EO data proxies for these variables are sourced to explore the ecology of Lyme disease to improve predictions (Ogden *et al.*, 2006a, 2010; Gabriele-Rivet *et al.*, 2015; Cheng *et al.*, 2017), to interpret surveillance data (Ogden *et al.*, 2006b, 2010), and ultimately to develop risk maps. Forest landscape patterns (including connectivity and fragmentation) in Canada have been mapped using EO data at a range of spatial resolutions (Wulder *et al.*, 2008; Pelletier *et al.*, 2017). Tick survival varies among woodland types (Guerra *et al.*, 2002; Ogden *et al.*, 2006a) so these patterns may add value to the assessment of the spatial distribution of Lyme disease risk. Assessments of Lyme disease risk across a range of geographic extents are needed to inform the different levels of jurisdiction (from national to regional and local) and help drive the design of their public health policies and programs.

The detail of LULC maps now obtainable from high-spatial-resolution satellite imagery (Fig. 2.2.1) means that EO data can serve assessment of risk for all levels of jurisdiction. Using spatial modeling techniques to produce maps of Lyme risk, the European LymeMAP project[1] combined data from a variety of sources. These include EO satellite data, terrestrial observations, Lyme disease data on humans from diagnostic laboratories, data from general practitioners, and data from end users such as health care organizations and professionals, resident, and visiting populations. Similar multi-data approaches are used to identify risk in Canada.[2]

Challenges and questions

There are two main challenges associated with the effective utilization of EO data for Lyme disease research and risk mapping efforts. The first relates to the identification of suitable EO-derived information sources that can help identify habitats suitable for ticks and their hosts. While generalized identification of woodland habitats may be adequate under some circumstances, more detailed assessment is needed by those responsible for managing Lyme disease risk at a local level of jurisdiction. This includes identification of variations in different woodland types regarding their suitability for ticks (Ogden *et al.*, 2006b) and for different communities of hosts, which may impact the abundance and the species and strains of Lyme disease-causing bacteria (Kurtenbach *et al.*, 2006; Mechai *et al.*, 2016). Consequently, EO data fine enough in spatial resolution and capacity to resolve different woodland types are needed. The second challenge relates to the selection and access to EO sources of climate data that are useful for mapping the distributions of the vectors and hosts under current and future climate using climate model output.

There are several questions and issues to be answered and resolved, including:

- What EO data can be used as proxies for key habitat/environmental indicators for ticks, hosts, and pathogens at a range of geographic scales from regional to local?
- What EO data are appropriate climate indicators at a range of geographic scales from regional to local?
- What EO data and products can be used to assist prediction of the expansion of Lyme disease risk areas?
- What future EO instruments might be able to provide tools and geospatial data to better understand the impacts of habitat, biodiversity, and climate change on tick-borne diseases such as Lyme disease?

Responses and options

Below are responses of the experts consulted regarding key issues and the data needed for public health responses to tick-borne diseases:

- EO and geospatial data are needed at various geographic scales depending on their use. The data required are environmental variables affecting distribution and activity of vectors and pathogens, and include land cover, land use, forest fragmentation and type, temperature and precipitation, surface, and soil moisture, and perhaps snow cover in some locations.
- Suitable non-commercial EO data include those from Landsat, Sentinel, and RADARSAT Constellation Mission (RCM) for land use and land cover variables (forest types, forest fragments, etc.) and those from

Fig. 2.2.1. (A) High-resolution satellite image suitable for extracting key land use information, e.g. woodlands and edges; (B) Map of tick habitats in a parkland area near Bristol, UK, used to analyze dog walking routes and tick exposure. (From: Jennett *et al.*, 2013.)

VIIRS, MODIS, AVHRR, and 3B42/TMPA (TRMM continuity) for climatic and microclimatic variables (surface temperature, surface moisture, air temperature, air humidity, precipitation, etc.).
- Required EO data sets are generally available either online, or in the Canadian context, through various government departments.
- Required spatial and temporal details of EO data are generally in the order of a few tens or hundreds of meters, whereas the revisit periods for mapping are in the order of one to five years.

EO-based risk maps for Lyme disease in central and eastern Canada

Data on risk factors used in risk assessment models of Lyme disease focus on favorable climatic conditions and suitable habitats. Model input is often based on coarse spatial resolution information. However, fine spatial resolution data are required to capture landscape and habitat heterogeneity and microclimate variations for risk prediction for Lyme disease at local scales. Local-scale risk prediction is needed for local-level management, which includes implementation of surveillance, prevention, and control measures such as warning signs and landscape management in parks and gardens (Stafford, 2007). To meet this need, local characterization of Lyme disease risk may be achieved by an indicator-based approach using high-spatial-resolution EO data. In contrast to mosquito-borne diseases, for which the time scale of vector generation times and transmission cycles may be a few weeks, the time scale for risk maps based on EO data needed for tick-borne diseases is in the order of 2–3 years or more. This time frame corresponds to the lifecycle of the tick over which environmental conditions must remain suitable for tick populations to survive.

Risk maps for central and eastern Canada with high spatial resolution using EO data on occurrence of suitable habitat and suitable climate have been developed (Kotchi *et al.*, 2021). An example of maps for possible occurrence of *I. scapularis* tick populations – and by inference, Lyme disease risk – were produced using EO data (Fig. 2.2.2). The map estimates risk using land cover and climate data from 2000 to 2015. The risk maps were validated with tick surveillance data.

Expected outcomes and impacts

What does this map do? The *I. scapularis* risk map identifies areas where the environment is suitable for populations of this tick to become established in eastern Canada. In northeastern North America, including Canada, invasion of this tick is accompanied by invasion of the pathogens of zoonotic infections transmitted by the tick. Therefore, identifying where tick populations could be also identifies where there is risk of acquiring Lyme disease and other *I. scapularis*-borne pathogens (*Babesia microti*, *Anaplasma phagocytophilum*, *Borrelia miyamotoi*, *B. mayonii*, Powassan and deer tick viruses). Knowing where this risk is at present or in the near future is imperative for public health responses.

The first public health response that these maps would initiate is surveillance. The risk maps identify where risk *might* be, but surveillance is needed to identify where risk actually *is*. At and ahead of the leading edge of tick population spread, many environmentally suitable niches have not yet been filled by tick populations (Clow *et al.*, 2017; Gabriele-Rivet *et al.*, 2017). Identifying emerging areas of risk is a constant preoccupation of local public health organizations because identifying risk is the starting point for prevention initiatives.[3] In areas known for recent and widespread in-migration of ticks, risk maps may well identify with sufficient accuracy where current risks are, precluding the need for further surveillance.

The risk map can be updated annually and show how risk may vary from one year to the next due to variations in ambient temperature over the 2–3-year lifecycle of the tick. It is also possible to track changes in risk associated with a warming climate over time scales of 5–10 years; and forecast each year whether risk is likely to be particularly high (or not) due to temperature conditions over the preceding years.

Why use EO data? By using EO data, risk can be identified anywhere in central and eastern Canada, including areas where ground-level data on environmental suitability are limited or

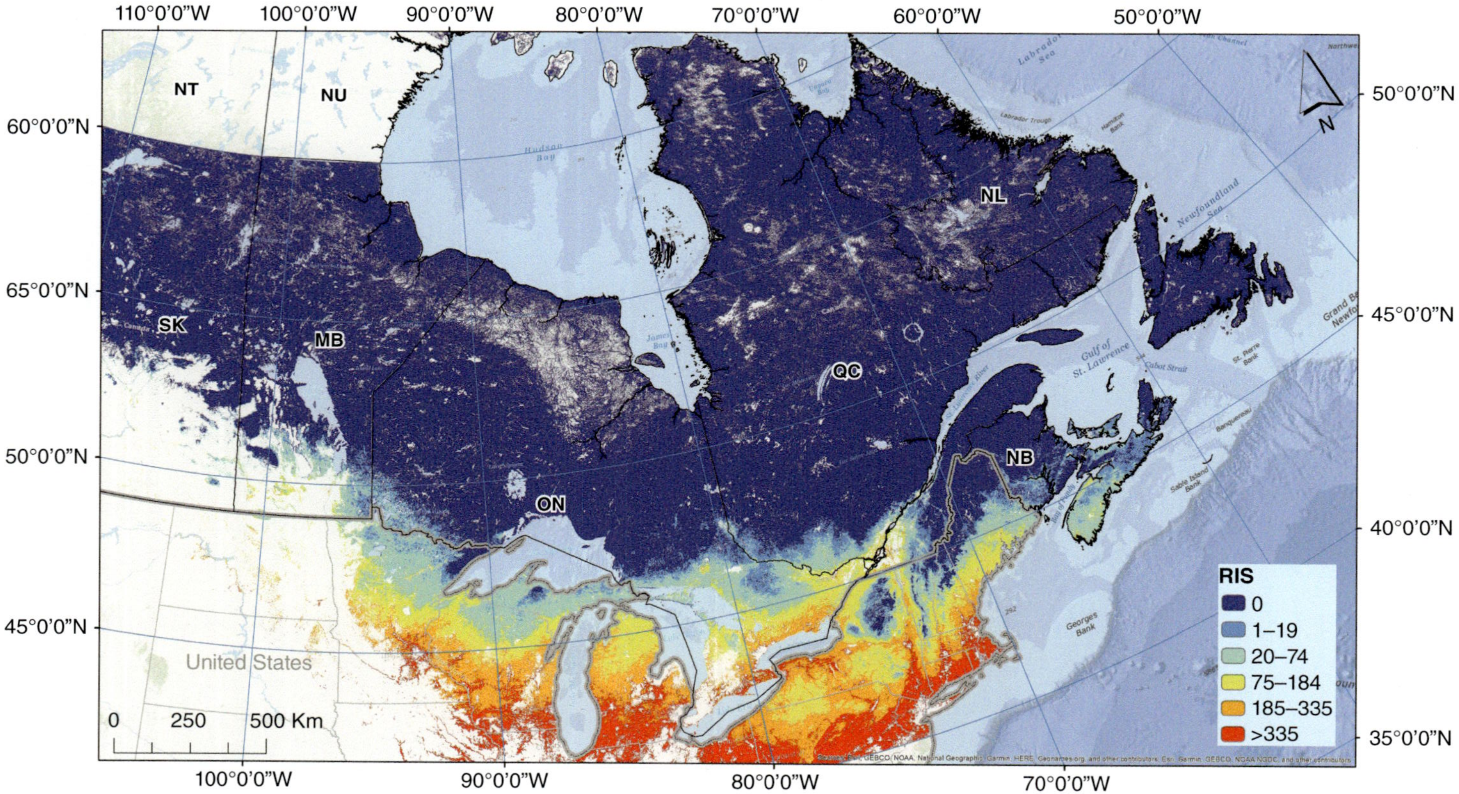

Fig. 2.2.2. Example of a risk map for *Ixodes scapularis* ticks (and by inference Lyme disease risk) in central and eastern Canada combining climatic and habitat suitability from 2000 to 2015. Color represents the levels of risk occurrence for ticks (RIS, risk of *I. scapularis*). Gray indicates areas with no ecological niches for *I. scapularis* ticks. (From: Kotchi *et al*., 2021.)

Table 2.2.1. Lyme risk map data needs and constraints related to their use.

Data	Type of geospatial data	Spatial resolution (SR)	Temporal resolution (TR)	Data availability	Complexity of EO data analysis (low/med/high)	EO data cost (free, nominal, significant)	Data source
Land use and land cover (LULC)	1. Product derived from Earth Observation (EO) images 2. Classification of EO images	1. SR ≤500 m 2. SR ≤30 m	1. Yearly 2. Quarterly	1. Available 2. Not available for the combined expected spatial and temporal resolutions	1. Low 2. High	Free	1. MODIS, VIIRS 2. Landsat-8, RapidEye, ACI, LCC
Forest type (FT)	Estimated from LULC images	SR ≤30 m	Quarterly	Processing possible (data available)	Medium	Free	Landsat-8, RapidEye, NFI, ACI, LCC
Forest fragmentation index (FFI)	Estimated from LULC images	SR ≤30 m	Quarterly	Not available	Medium	Free	Landsat-8, RapidEye, NFI, ACI, LCC
Land surface temperature (LST)	1. Product derived from EO images 2. Estimated using products derived from EO images and downscaling algorithms	As fine as possible: 1. SR ≤1000 m 2. SR ≤30 m	Daily	Available for option 1 Not for option 2 with the expected TR	1. Medium 2. Medium	Free	1. MODIS, AVHRR 2. MODIS & Landsat-8
Accumulated surface degree-days	Estimated using LST	As fine as possible: 1. SR ≤1000 m 2. SR ≤30 m	Daily	Available for option 1 Not for option 2 with the expected TR	Medium	Free	1. MODIS, AVHRR 2. MODIS & Landsat-8
Surface/soil moisture	Estimated using EO images	As fine as possible: 1. SR ≤1000 m 2. SR ≤30 m	Daily	Available for option 1 Not for option 2 with the expected TR	High	1. Free 2. Nominal	1. SMAP 2 RADARSAT-2

For forests, it is often enough to identify seasonal variations in forest parameters (for classification of forest types) and identify where forests have been cut down. For temperature, it is often enough to be able to obtain proxies for annual cumulative degree-days. For soil moisture, at least seasonal variations are needed.

ACI, Annual Crop Inventory (Agriculture and Agri-Food Canada); AVHRR, advanced very high-resolution radiometer; LCC, 2005–2010 20 m land cover of Canada (Natural Resources Canada); MODIS, Moderate Resolution Imaging Spectroradiometer; NFI, National Forest Inventory (Natural Resources Canada); SMAP 2, Soil Moisture Active Passive ; VIIRS, Visible/Infrared Imager Radiometer Suite.

absent. Currently available EO data integrated into the map would allow Lyme risk areas to be identified at a much finer spatial resolution than maps developed previously. The spatial detail afforded by the user of EO data would therefore serve the information needs of multiple end users, as it can identify risk nationally and provincially, and at a scale useful for municipalities and the public.

Who are the end users? Federal public health and health organizations need to identify at a national scale where risk is or may be occurring and where national surveillance prevention and control would be best targeted. The map may be used for national forecasting of high-risk years. It identifies and practicably tracks impacts of climate change on health in the context of Lyme disease. Provincial public health organizations and municipalities can identify at practical scale levels where risk is or may be occurring, helping to identify where surveillance, prevention, and control efforts would be best targeted. The map may be used for provincial forecasting of high-risk years. In regions where the tick populations are known to have become established in most suitable environments, the map can be used (at higher magnification) to inform the public of where risk occurs and where personal protection and prevention should be undertaken.

Technical considerations and perspectives for producing risk maps

This section highlights the technical considerations necessary for the use of EO data in the production of risk maps for tick-borne diseases.

What EO data are needed? The production of risk maps of the expansion of *I. scapularis* in central and eastern Canada on an annual basis and at a spatial resolution of 500 m requires computation of a risk model using surface temperature, surface reflectance, land use, and land cover products (Table 2.2.1).

What resources are needed? Infrastructure for big data storage: Even though the lifecycle of the tick is multiple years, production of risk maps on an annual basis and at a spatial resolution of 500 m requires the downloading of many EO images because very-high-temporal-resolution images are needed to obtain meaningful averages for microclimatic conditions important for *I. scapularis* ticks (Kotchi *et al.*, 2021). There is a requirement for higher spatial resolution (≤30 m) because Lyme disease risk can occur in small woodland patches (McClure and Diuk-Wasser, 2018), which would significantly increase the need for data storage. The processing of these data and the calculation of the variables needed to produce the risk map generates additional images, all of which require big data storage and analysis.

Infrastructure for big data processing: Image processing and analysis can be challenging. The number of images to be processed and the number of variables explored require the development of automated processes and related algorithms (Hermosilla *et al.*, 2016). Image analysis represents big data processing and requires high-quality computing capacity and availability for many days to run the programs.

Software and remote sensing skills: The analysis of the images and the production of the risk map require image processing software and a GIS. Expertise in EO image analysis, geo-informatics, and mapping is essential for the development of risk maps.

What future developments are needed? Validation using surveillance data is needed to assess the performance of the risk maps, and these surveillance data may be used iteratively to improve the accuracy of the risk model by suggesting alterations to the algorithm.

In order to support disease management at a local scale, it would be useful to produce risk maps at a higher spatial resolution (30 m). This would better assess risk at smaller spatial extents such as parks and woodlots within mixed environments such as urban agglomerations. Prevention and control actions at the local level may then better target these areas. Producing and archiving maps with high spatial resolution (30 m) would require exceptional data storage capacity.

What are the opportunities with EO products and data?

The availability of EO images, products derived from these images, and other geospatial data from national geodatabases offers several opportunities for the development of risk maps with high spatial and temporal resolutions. Some opportunities can be considered:

- Recent (e.g. Landsat-8, Sentinel-1, Sentinel-2, Spot-7, Sentinel-3) and future developments in EO systems with improved spatial, temporal, and radiometric resolutions.
- EO data and products covering national and global scales with a high update frequency are available from different geospatial databases (e.g. Canadian Forest Service of Natural Resources Canada, Annual Crop Inventory from Agriculture and Agri-food Canada, etc.) and data portals (e.g. USGS EarthExplorer, NASA Earth Observing System Data and Information System, Google Earth Engine, etc.).
- Downscaling approaches based on multi-sensor data integrating thermal images with very high temporal resolution (daily) and multispectral images with high spatial resolution would provide the opportunity to obtain surface temperature data with both high spatial and temporal resolutions.
- Following investigation, data from new EO systems such as the RADARSAT Constellation Mission may be used to directly estimate microclimatic variables like soil and surface moisture both at very high spatial and temporal resolution.

Current products and developments in EO that contribute to the public health sector

The factors governing the establishment of tick vector populations for Lyme disease are associated with their habitat (e.g. presence of host animals and refuges to survive winter) and favorable climatic conditions over the length of the lifecycle. Tick habitat is mainly broadleaf or mixed-wood forest that provides a leaf litter layer sufficient to provide a refuge for ticks during adverse (cold, hot, or dry) weather. To identify suitable forests and forest fragments, the Canadian Forest Service of the Department of Natural Resources Canada offers a multitude of products with 30 m or 250 m resolution generated through ongoing research activities, as follows.

Land cover and land cover change: This product is a critical source of information for forest habitat characterization and monitoring and biodiversity assessment. A time series-based, disturbance-informed, land cover mapping approach is demonstrated over the ~650 Mha of forested ecosystems in Canada (Hermosilla *et al.*, 2018). The production of land cover is enabled by local high-performance computing for initial data preparation and access to a supercomputing environment for spectral classification, incorporation of change, and successional logic, which results in an integrated 28-year data cube of land cover (Wulder *et al.*, 2018). The processing framework used to generate this product is called the Virtual Land Cover Engine, or VLCE, which can incorporate any training data and generate a national time series of land cover according to a specified land cover legend. VLCE land cover products for Canada have subsequently been used to identify areas of forest land use in keeping with national and international reporting requirements (Wulder *et al.*, 2020a).

Stand replacing forest disturbance: This product is a national Landsat time series product characterizing forest disturbance (White *et al.*, 2017). It incorporates a change hierarchy for stand replacing and non-stand replacing changes (Hermosilla *et al.*, 2015b) and characterizes multiple changes at the same location over time (Hermosilla *et al.*, 2019). This product could provide valuable information about vector habitat disturbance.

Forest structure: This product informs on forest structural attributes like canopy cover, height, volume, and total aboveground biomass (Zald *et al.*, 2016; Matasci *et al.*, 2018a). These attributes could be potential proxy variables to assess habitat suitability for vectors. The forest structure product is based on Landsat time series and national light detection and ranging (LiDAR) transect data. Three decades of forest structure information for Canada's forested ecosystems (1985–2016) have been generated (Matasci *et al.*, 2018b) and have subsequently been used to characterize long-term trends in treed aboveground biomass dynamics (Wulder *et al.*, 2020b).

Best-available pixel composite: This is a product that provides national, annual, cloud-free, surface reflectance image composites derived from Landsat data using the Composite to Change (C2C) protocol. These data products provide opportunities to derive other information products and Essential Climate Variables (ECVs) (e.g. vegetation indices) (White *et al.*, 2014; Hermosilla *et al.*, 2015a, 2016). This product could be very useful not only for rapid assessment of Lyme disease risk but also for many other vector-borne diseases including mosquito-borne diseases.

Forest fragmentation time series: A product was developed for the ~650 million hectares of Canada's forested ecosystems (Hermosilla *et al.*, 2018), building on work by Wulder *et al.* (2008, 2011) and by Pelletier *et al.* (2017). These data could provide important information for the local characterization of Lyme disease risk and could also support prediction of the spread of tick and tick-borne pathogen populations from one location to another.

Maps of Canada's forest attributes for 2001 and 2011 at 250 m resolution: This Canadian Forest Service map product, based on Canada's National Forest Inventory (Beaudoin *et al.*, 2017), includes percentage forest species composition for 73 species as well as land cover and forest structural attributes. The species composition could again be useful in identifying habitats suitable for ticks and the pathogens they transmit.[4]

In agricultural regions and urban agglomerations, forest fragments are part of the landscape but are generally not well characterized by geospatial products from the forest sector. Agriculture and Agri-Food Canada (Davidson *et al.*, 2017) complements the data available by providing products in which these forest fragments and other environmental determinants are identified, as follows.

Land use (LU) maps: These maps are decadal products that cover all areas of Canada south of 60°N at a spatial resolution of 30 m. Maps were released in 1990, 2000, 2005, 2010, 2015, and pending 2020.[5] LU classes in the maps are based on the Intergovernmental Panel on Climate Change legend and include forest, water, cropland, grassland, settlement, and "other," which includes barren land, ice, rock, and unclassified land. Methods have been developed to harmonize land cover products, such as annual crop inventory (ACI) and VLCE, which could be very useful in generating wall-to-wall land cover products for Canada's terrestrial area (Li *et al.*, 2021).

ACI maps: These maps have been available since 2009 for the Canadian prairies and 2011 nationally. They are also a 30-m resolution product that covers all the Canadian provinces. They are based on a decision tree model using optical (Advanced Wide Field Sensors [AWiFs], Landsat-5, 7, and 8, Sentinel-2, Gaofen-1) and SAR (Sentinel-1, RADARSAT-2) satellite images. The ACI land use and land cover classes include surface water, urban and developed areas, shrublands, wetlands, grasslands, agriculture (including a range of crop varieties), and forest types (coniferous, broadleaf, mixed wood).

The 2005–2010 20 m land cover of Canada south of tree-line product from the Canada Centre for Mapping and Earth Observation, Natural Resources Canada, is another land use and land cover product that complements the ones above. It is based on satellite images from the Système Pour l'Observation de la Terre (SPOT-4 and SPOT-5) satellites. It consists of 16 generic classes including more than five forest types (evergreen conifer forest, evergreen conifer forest, deciduous forest, mixed forest, etc.), wetlands, and surface waters.

Findings and opportunities for collaboration

Emerging infectious diseases are a significant public health issue for Canada. Geospatial data that identify the drivers of disease emergence or the factors that influence disease endemicity must be available to the public health community. The example of the *I. scapularis* risk map shows that geospatial data producers in the Canadian federal community provide reference data of the quality and relevance needed to feed predictive models of vector-borne disease and risk maps. These data have the potential to refine risk models by offering a more detailed assessment of their capacity to identify environmental suitability for different tick species and tick-borne pathogens, and by providing a range of spatial and temporal resolutions that can help multiple jurisdictions in risk communication and disease management.

The data offered by the different EO and geomatics sectors, together with the ongoing research efforts of public health scientists in the field of tick-borne diseases, provide the basis for a network for collaboration. Also, the infrastructure, technologies, methods, and supercomputing environment developed by departments involved in EO and geomatics should serve as an example for the establishment of an infrastructure framework to support the development of geomatics in the public health sector.

Notes

[1] https://business.esa.int/projects/lymemap (accessed 3 January 2022).
[2] https://www.canada.ca/en/public-health/services/diseases/lyme-disease/risk-lyme-disease.html (accessed 3 January 2022).
[3] https://www.inspq.qc.ca/zoonoses/maladie-de-lyme (accessed 3 January 2022).
[4] https://open.canada.ca/data/en/dataset/ec9e2659-1c29-4ddb-87a2-6aced147a990 (accessed 3 January 2022).
[5] https://open.canada.ca/data/en/dataset/fa84a70f-03ad-4946-b0f8-a3b481dd5248 and https://open.canada.ca/data/en/dataset/fa84a70f-03ad-4946-b0f8-a3b481dd5248/resource/7575a7a5-4d28-478c-b32f-e9ae7c15c622 (both accessed 19 January 2022).

References

Aenishaenslin, C., Bouchard, C., Koffi, J.K., and Ogden, N.H. (2017) Exposure and preventive behaviours toward ticks and Lyme disease in Canada: Results from a first national survey. *Ticks and Tick-borne Diseases* 8, 112–118. DOI: https://doi.org/10.1016/j.ttbdis.2016.10.006.

Beaudoin, A., Bernier, P.Y., Villemaire, P., Guindon, L., and Jing Guo, X. (2017) Tracking forest attributes across Canada between 2001 and 2011 using a *k* nearest neighbors mapping approach applied to MODIS imagery. *Canadian Journal of Forest Research* 48(1) 85–93. DOI: https://doi.org/10.1139/cjfr-2017-0184.

Bouchard, C., Leonard, E., Koffi, J.K., Pelcat, Y., Peregrine, A. *et al.* (2015) The increasing risk of Lyme disease in Canada. *Canadian Veterinary Journal* 56, 693–699. Available at: https://www.ncbi.nlm.nih.gov/pmc/articles/PMC4466818/ (accessed 5 August 2021).

Bouchard, C., Aenishaenslin, C., Rees, E.E., Koffi, J.K., Pelcat, Y. *et al.* (2018) Integrated social-behavioral and ecological risk maps to prioritize local public health responses to Lyme disease. *Environmental Health Perspectives* 126. DOI: https://doi.org/10.1289/EHP1943.

Cheng, A., Chen, D., Woodstock, K., Ogden, N.H., Wu, X. *et al.* (2017) Analyzing the potential risk of climate change on Lyme disease in Eastern Ontario, Canada using time series remotely sensed temperature data and tick population modelling. *Remote Sensing* 9, 609. DOI: https://doi.org/10.3390/rs9060609.

Clow, K., Leighton, P.A., Ogden, N.H., Lindsay, L.R., Michel, P. *et al.* (2017) Northward range expansion of *Ixodes scapularis* evident over a short timescale in Ontario, *Canada. PLoS ONE* 12. DOI: https://doi.org/10.1371/journal.pone.0189393.

Davidson, A.M., Fisette, T. McNairn, H. and Daneshfar, B. (2017) Detailed crop mapping using remote sensing data (crop data layers). In: Delincé, J. (ed.) *Handbook on Remote Sensing for Agricultural Statistics*. GSARS, Rome. Available at: http://www.fao.org/3/ca6394en/ca6394en.pdf (accessed 14 June 2021).

Gabriele-Rivet, V., Arsenault, J., Badcock, J., Cheng, A., Edsall, J. *et al.* (2015) Different ecological niches for ticks of public health significance in Canada. *PLoS ONE* 10. DOI: https://doi.org/10.1371/journal.pone.0131282.

Gabriele-Rivet, V., Koffi, J.K., Pelcat, Y., Arsenault, J., Cheng, A. *et al.* (2017) A risk model for the Lyme disease vector *Ixodes scapularis* (Acari: Ixodidae) in the Prairie Provinces of Canada. *Journal of Medical Entomology* 54, 862–868. DOI: https://doi.org/10.1093/jme/tjx036.

Guerra, M., Walker, E., Jones, C., Paskewitz, S., Cortinas, M.R. *et al.* (2002) Predicting the risk of Lyme disease: habitat suitability for *Ixodes scapularis* in the north central United States. *Emerging Infectious Diseases* 8, 289–297. Available at: https://stacks.cdc.gov/view/cdc/13770 (accessed 5 August 2021).

Hermosilla, T., Wulder, M.A., White, J.C., Coops, N.C., and Hobart, G.W. (2015a) An integrated Landsat time series protocol for change detection and generation of annual gap free surface reflectance composites. *Remote Sensing of Environment* 158, 220–234. DOI: https://doi.org/10.1080/17538947.2016.1187673.

Hermosilla, T., Wulder, M.A., White, J.C., Coops, N.C., and Hobart, G.W. (2015b) Regional detection, characterization, and attribution of annual forest change from 1984 to 2012 using Landsat-derived time-series metrics. *Remote Sensing of Environment* 170, 121–132. DOI: https://doi.org/10.1016/j.rse.2015.09.004.

Hermosilla, T., Wulder, M.A., White, J.C., Coops, N.C., Hobart, G.W. *et al.* (2016) Mass data processing of time series Landsat imagery: Pixels to data products. *International Journal of Digital Earth* 9, 1035–1054. DOI: https://doi.org/10.1080/17538947.2016.1187673.

Hermosilla, T., Wulder, M.A., White, J.C., Coops, N.C., and Hobart, G.W. (2018) Disturbance-informed annual land cover classification maps of Canada's forested ecosystems for a 29-year Landsat time series. *Canadian Journal of Remote Sensing* 44, 67–87. https://doi.org/10.1080/07038992.2018.1437719.

Hermosilla, T., Wulder, M.A., White, J.C., Coops, N.C., Pickell, P.D. *et al.* (2019) Impact of time on interpretations of forest fragmentation: Three-decades of fragmentation dynamics over Canada. *Remote Sensing of Environment* 222, 65–77. DOI: https://doi.org/10.1016/j.rse.2018.12.027.

Jennett, A.L., Smith, F.D., and Wall, R. (2013) Tick infestation risk for dogs in a peri-urban park. *Parasites & Vectors* 6, 358. DOI: http://doi.org/10.1186/1756-3305-6-358.

Kotchi, S.O., Bouchard, C., Brazeau, S., and Ogden, N.H. (2021) Earth observation-informed risk maps of the Lyme disease vector *Ixodes scapularis* in central and eastern Canada. *Remote Sensing* 13, 524. DOI: https://doi.org/10.3390/rs13030524.

Kurtenbach, K., Hanincová, K., Tsao, J., Margos, G., Fish, D. *et al.* (2006) Fundamental processes in the evolutionary ecology of Lyme borreliosis. *Nature Reviews Microbiology* 4, 660–669. DOI: https://doi.org/10.1038/nrmicro1475.

Leighton, P.A., Koffi, J.K., Pelcat, Y., Lindsay, L.R., and Ogden, N.H. (2012) Predicting the speed of tick invasion: An empirical model of range expansion for the Lyme disease vector *Ixodes scapularis* in Canada. *Journal of Applied Ecology* 49, 457–464. DOI: https://doi.org/10.1111/j.1365-2664.2012.02112.x.

Li, Z., White, J.C., Wulder, M.A., Hermosilla, T., Davidson, A.M. *et al.* (2021) Land cover harmonization using Latent Dirchlect Allocation. *International Journal of Geographical Information Science* 35, 348–374. DOI: https://doi.org/10.1080/13658816.2020.1796131.

Matasci, G., Hermosilla, T., Wulder, M.A., White, J.C., Coops, N.C. *et al.* (2018a) Large-area mapping of Canadian boreal forest cover, height, biomass and other structural attributes using Landsat composites and lidar plots. *Remote Sensing of Environment* 209, 90–106. DOI: https://doi.org/10.1016/j.rse.2017.12.020.

Matasci, G., Hermosilla, T., Wulder, M.A., White, J.C., Coops, N.C. *et al.* (2018b) Three decades of forest structural dynamics over Canada's forested ecosystems using Landsat time-series and lidar plots. *Remote Sensing of Environment* 216, 697–714. DOI: https://doi.org/10.1016/j.rse.2018.07.024.

McClure, M. and Diuk-Wasser M. (2018) Reconciling the entomological hazard and disease risk in the Lyme disease system. *International Journal of Environmental Research and Public Health* 15. DOI: https://doi.org/10.3390/ijerph15051048.

Mechai, S., Margos, G., Feil, E.J., Barairo, N., Lindsay, L.R., *et al.* (2016) Evidence for host-genotype associations of *Borrelia burgdorferi* sensu stricto. *PLoS ONE* 11. DOI: https://doi.org/10.1371/journal.pone.0149345.

Ogden, N.H. and Lindsay, L.R. (2016) Effects of climate and climate change on vectors and vector-borne diseases: Ticks are different. *Trends in Parasitology* 32, 646–656. DOI: https://doi.org/10.1016/j.pt.2016.04.015.

Ogden, N.H., Barker, I.K., Beauchamp, G., Brazeau, S., Charron, D. *et al.* (2006a) Investigation of ground level and remote-sensed data for habitat classification and prediction of survival of *Ixodes scapularis* ticks in habitats of southeastern Canada. *Journal of Medical Entomology* 43, 403–414. Available at: https://www.researchgate.net/publication/7159896_Investigation_of_Ground_Level_and_Remote-Sensed_Data_for_Habitat_Classification_and_Prediction_of_Survival_of_Ixodes_scapularis_in_Habitats_of_Southeastern_Canada (accessed 15 June 2021).

Ogden, N.H., Trudel, L., Artsob, H., Barker, I.K., Beauchamp, G. *et al.* (2006b) *Ixodes scapularis* ticks collected by passive surveillance in Canada: Analysis of geographic distribution and infection with the Lyme borreliosis agent *Borrelia burgdorferi*. *Journal of Medical Entomology* 43, 600–609. DOI: https://doi.org/10.1093/jmedent/43.3.600.

Ogden, N.H., St-Onge, L., Barker, I.K., Brazeau, S., Bigras-Poulin, M., *et al.* (2008) Risk maps for range expansion of the Lyme disease vector, *Ixodes scapularis*, in Canada now and with climate change. *International Journal of Health Geographics* 7, 24. DOI: https://doi.org/10.1186/1476-072X-7-24.

Ogden, N.H., Bouchard, C., and Kurtenbach, K. (2010) Active and passive surveillance and phylogenetic analysis of *Borrelia burgdorferi* elucidate the process of Lyme disease risk emergence in Canada. *Environmental Health Perspectives* 118, 909–914. DOI: https://doi.org/10.1289/ehp.0901766.

Ogden, N.H., Radojević, M., Wu, X., Duvvuri, V.R., Leighton, P.A. *et al.* (2014) Estimated effects of projected climate change on the basic reproductive number of the Lyme disease vector *Ixodes scapularis*. *Environmental Health Perspectives* 122, 631–638. DOI: http://dx.doi.org/10.1289/ehp.1307799.

Pelletier, D., Lapointe M.É., Wulder MA., White J.C., and Cardille J.A. (2017) Forest connectivity regions of Canada using circuit theory and image analysis. *PLoS ONE* 12. DOI: https://doi.org/10.1371/journal.pone.0169428.

Stafford, K.C. (2007) Tick management handbook: An integrated guide for homeowners, pest control operators, and public health officials for the prevention of tick-associated disease. *Bulletin Connecticut Agricultural Experiment Station* 1010, 78. Available at: https://portal.ct.gov/-/media/caes/documents/publications/bulletins/b1010pdf (accessed 15 June 2021).

White, J.C., Wulder, M.A., Hobart, G.W., Luther, J.E., Hermosilla, T. *et al.* (2014) Pixel-based image compositing for large-area dense time series applications and science. *Canadian Journal of Remote Sensing* 40, 192–212. DOI: https://doi.org/10.1080/07038992.2014.945827.

White, J.C., Wulder, M.A., Hermosilla, T., Coops, N.C., and Hobart, G. (2017) A nationwide annual characterization of 25 years of forest disturbance and recovery for Canada using Landsat time series. *Remote Sensing of Environment* 194, 303–321. DOI: https://doi.org/10.1016/j.rse.2017.03.035.

Wulder, M.A., White J.C., Han T., Coops N.C., Cardille J.A. *et al.* (2008) Monitoring Canada's forests. Part 2: National forest fragmentation and pattern. *Canadian Journal of Remote Sensing* 34, 563–584. DOI: https://doi.org/10.5589/m08-081.

Wulder, M.A., White, J.C., and Coops, N.C. (2011) Fragmentation regimes of Canada's forests. *The Canadian Geographer* 55, 288–300. DOI: https://doi.org/10.1111/j.1541 0064.2010.00335.x.

Wulder, M.A., Coops, N.C., Roy, D.P., White, J.C., and Hermosilla, T. (2018) Land cover 2.0. *International Journal of Remote Sensing* 39, 4254–4284. DOI: https://doi.org/10.1080/01431161.2018.1452075.

Wulder, M.A., Hermosilla, T., Stinson, G., Gougeon, F.A., White, J.C. *et al.* (2020a) Satellite-based time series land cover and change information to map forest area consistent with national and international reporting requirements. *Forestry: An International Journal of Forest Research* 93, 331–343. DOI: https://doi.org/10.1093/forestry/cpaa006.

Wulder, M.A., Hermosilla, T., White, J.C., and Coops, N.C. (2020b) Biomass status and dynamics over Canada's forests: Disentangling disturbed area from associated aboveground biomass consequences. *Environmental Research Letters* 15, 094093. DOI: https://doi.org/10.1088/1748-9326/ab8b11.

Zald, H.S.J., Wulder, M.A., White, J.C., Hilker, T., Hermosilla, T. *et al.* (2016) Integrating Landsat pixel composites and change metrics with lidar plots to predictively map forest structure and aboveground biomass in Saskatchewan, Canada. *Remote Sensing of Environment* 176, 188–201. DOI: https://doi.org/10.1016/j.rse.2016.01.015.

2.3 Air Quality and Heat-related Health Issues

John Haynes[1], Didier Davignon[2], Jack Chen[3], Yves Rochon[3], Chris McLinden[3], Christopher Sioris[3], and Dirk Werle[4]

[1]Health and Air Quality Applications, Applied Sciences Program, NASA Earth Science Division, USA; [2]Canadian Meteorological Centre Operations, Environment and Climate Change Canada, Montréal, Canada; [3]Air Quality Research Division, Environment and Climate Change Canada, Montréal, Canada; [4]Ærde Environmental Research, Halifax, Canada

Context, state of knowledge, challenges, and responses

Epidemiological evidence suggests that air pollution has become the biggest environmental cause of premature death, overtaking poor sanitation and a lack of clean drinking water (Organisation for Economic Co-operation and Development [OECD], 2014). Two additional seasonal environmental issues are of importance globally, as well as in Canada: the effect of wildfire smoke plumes on air pollution (Sofowote and Dempsey, 2015; Le *et al.*, 2014) and the effect of urban heat islands on the health of urban populations (Giguère, 2009). One of the challenges for public health authorities is to mitigate the impact of air pollution and urban heat islands on vulnerable groups, such as the very young and the elderly.

Environment and Climate Change Canada (ECCC) provides an air quality forecast service as part of its nationwide online weather information network.[1] The ECCC air quality forecast program identifies three dominant indicators of summer and winter smog, which are ground-level ozone (O_3), nitrogen dioxide (NO_2), and fine particulate matter (PM) that is 2.5 microns and less in diameter ($PM_{2.5}$). They were selected based on the strength of evidence of their impact on health, and their respective concentration–response ratios were statistically combined to produce the Air Quality Health Index (AQHI) (Stieb *et al.*, 2008). A key supporting component of the AQHI program is the Canadian surface measurement network, which continuously monitors ozone, $PM_{2.5}$, and NO_2 data in populated areas. In the USA, the National Oceanic and Atmospheric Administration (NOAA) develops and implements operational air quality forecast guidance, and the Environmental Protection Agency (EPA) produces an air quality index (AQI) forecast based on pollutant standards they have set. The US AQI is a standards-based worst-pollutant index that incorporates five major pollutants and is supported by a large monitoring network reporting under the AirNow data management and dissemination tool (e.i. AirNow example p.44).

In recent years, measurements of thermal conditions in urban areas have allowed for better characterization of urban heat island effects. This has led to the modernization of heat alert and response programs to counter the adverse heat and air quality impacts on human health, particularly among children, seniors, and people at greater risk of cardiovascular and pulmonary disease (De Sario *et al.*, 2013).

In addition to complementing *in situ* air quality monitoring at the surface, space-based EO data are employed to routinely support a variety of related measurement and mapping activities. They include estimating emissions, tracking pollutant plumes, supporting air quality forecasting activities, detecting wildland fires, providing evidence for "exceptional event" declarations, monitoring regional long-term trends, and evaluating air quality model output (Duncan *et al.*, 2014). Specialized EO sensor systems involve the Ozone Monitoring Instrument (OMI) on board NASA Aura satellite, the MODIS on board NASA's Terra and Aqua, and the Tropospheric Monitoring Instrument (TROPOMI). New satellite-based instruments are expected to be launched in the coming years, such as NASA's Tropospheric Emissions: Monitoring of Pollution (TEMPO) spectrometer onboard a geostationary satellite (planned for 2022).[2] TEMPO will make accurate hourly daytime measurements of tropospheric ozone, nitrogen dioxide, and formaldehyde, with significantly improved spatial resolution over other sensors currently available. Together with the Korean Geostationary Environment Monitoring Spectrometer (GEMS) and the ESA Copernicus Sentinal-4 missions, TEMPO will comprise the North American element of a geostationary satellite constellation to monitor air quality (Chance *et al.*, 2013; Zoogman *et al.*, 2017).

Examples of recent research

In general, air quality and weather forecast programs are large consumers of EO data. There are several areas of ongoing research with particular application in public health. They include the assimilation of real-time EO data into air quality forecast models; the use of near-real-time wildland fire emission estimates in smoke forecasts; and pollutant emission inference and hybrid approaches to surface pollutant exposure mapping. In order to capture local thermal anomalies, weather forecast modeling at the urban scale attempts to take advantage of detailed land use maps for developing appropriate urban surface models and assimilation systems.

One recent example of public health-related EO data assimilation is the use of real-time measurements for improving the Ultraviolet (UV) Index forecast. In the USA, an OMI surface UV irradiance product was spatially interpolated to the county level and matched with demographic and economic data available from the Centers for Disease Control and Prevention (CDC). This unique product provides a data source to map national distribution and long-term trends in UV radiation for risk communication and health-related studies. The project was a partnership between NASA and the CDC.[3]

Accurate characterization of the amount of stratospheric ozone allows for the correct forecast of the maximum amount of UV rays that will reach the Earth's surface. Overexposure to UV rays can cause a variety of health concerns ranging from sunburns to skin cancer. The UV Index forecasts serve as a guide to take protective steps for reducing or preventing overexposure to UV rays. The data assimilation system for weather forecasting has been expanded to use constituent data with the addition of stratospheric ozone measurements from different satellite instruments. These include the Global Ozone Monitoring Experiment-2 instruments (GOME-2A and GOME-2B) on the MetOp-A and MetOp-B satellites, the total column ozone mapping instrument (OMPS-NM) of the Ozone Mapping Profiler Suite on the NOAA 20 and Suomi National Polar-orbiting Partnership (NPP) satellites, the OMI on the Aura research satellite, TROPOMI, partial column ozone profiles from the Solar Backscatter Ultraviolet Radiometer instrument (SBUV/2) on the NOAA-19 satellite and OMPS-NP on the Suomi NPP satellite, and ozone profiles of the Microwave Limb Sounder (MLS) on the Aura satellite (e.g. Rochon *et al.*, 2019). The resulting ozone layer forecasts are leading to enhanced UV Index products, including tracking daytime changes and extending the forecast period (e.g. Tereszchuk *et al.*, 2018). Ozone data from other satellite instruments will be added over time as other sources become unavailable. These additional sources are expected to include TEMPO, as well as ozone-sensitive infrared spectral channels currently being investigated for this purpose from measurements of the Cross-track Infrared Sounders (CrIS) on Suomi NPP and NOAA-20, the Infrared Atmospheric Sounding Interferometer (IASI) on the MetOp satellite series, and the Atmospheric Infrared Sounder (AIRS) on the Aqua satellite.

There is considerable potential to combine ground-based and satellite-based observations in data assimilation systems for improving air quality forecast. Meteorological organizations are investing efforts in this direction. Current air quality forecast systems are limited by their ability to accurately capture the initial distribution of pollutants and to correctly define pollutant emission sources. The most significant gain in air quality forecast modeling consists of the incorporation of a data assimilation cycle with source inversion, where EO data can provide information on initial conditions and adjust emission rates based on recent observations (Peng *et al.*, 2017).

During the summer months, air quality and visibility degradation as a result of wildland fire smoke are becoming major concerns in many cities, especially those in western Canada and the USA. Wildland fires can contribute to large quantities of atmospheric pollutants and can increase the formation of secondary pollutants such as O_3 and fine aerosols. Moderate-resolution EO data from the polar-orbiting AVHRR, MODIS, and VIIRS satellite sensors are key components for identifying fire location and intensity in near real time, thus facilitating the quantification of emissions. Using a top-down approach with fire radiative power (FRP) as proxy for total fuel consumption, several global biomass burning emissions inventories are now available with varying spatial resolution and temporal coverage, as listed in Table 2.3.1. Some of these databases and inventories are being used in global chemical models (European Centre for Medium-Range Weather Forecasts – Monitoring Atmospheric Composition and Climate with the Global Fire Assimilation System [ECMWF-MACC with GFAS], NASA Goddard Chemistry Aerosol Radiation and Transport with Quick Fire Emissions Dataset [NASA-GOCART with QFED], National Center for Atmospheric Research – Whole Atmosphere Community Climate Model with Fire INventory from NCAR [NCAR-WACCM with FINN]) to simulate the long-range transport of smoke across continents.

The ECCC FireWork system utilizes near-real-time EO data and a chemical transport model to provide 48-h forecasts of air quality conditions across North America. FireWork is a valuable tool for regional air quality forecasters and emergency first responders for issuing health bulletins and evacuation warnings.

Air quality-related EO data has been utilized for deriving so-called "top-down" emission estimates for CO, NOx, SO_2, and volatile organic compounds, with some progress made with regard to $PM_{2.5}$ (Levelt *et al.*, 2018). The types of emissions captured by EO are varied and occur over different spatial and temporal scales; examples include natural forest fires, volcanic eruptions, and anthropogenic sources, as reviewed by Streets *et al.* (2013). OMI data have been used to pinpoint locations of SO_2 emissions and subsequently quantify their annual emissions by combining OMI observations with wind hind-cast information. A summary of over 500 locations is shown in Fig. 2.3.1 (Fioletov *et al.*, 2016); it includes several dozen locations of anthropogenic and volcanic sources not included in any other emissions inventories (McLinden *et al.*, 2016). This work resulted in an update to the Hemispheric Transport Air Pollutant (HTAP) SO_2 emissions inventory, referred to as OMI+HTAP, which led to improved prediction of SO_2 concentrations compared with surface networks (Liu *et al.*, 2018).

Satellite observations have also been used to determine long-term trends in ambient pollutant concentrations. Using OMI data, NASA established NO_2 emission trends for 195 cities worldwide.[4]

The use of EO data as a source of information concerning urban heat islands is well established. More specifically, satellite thermal instruments are used to characterize the intensity of surface urban heat islands, based on surface temperature as a proxy for air temperature. Recently, Chakraborty and Lee (2019) used the extensive archives of the MODIS instruments to create a global comprehensive characterization of surface urban heat islands (UHIs) across 9500 urban clusters using over 15 years of data, allowing for a classification of UHIs by climate zones and an estimation of trends. Canadian examples include the use of Landsat Thematic Mapper (TM) and Enhanced Thematic Mapper Plus (ETM+) to examine surface intra-urban heat islands in Montreal, QC (Martin *et al.*, 2015) (Fig. 2.3.2) and to study urban land surface temperature (LST) impacts in the city of Saskatoon, SK (Shen *et al.*, 2014). While the use of LST as a proxy for air surface temperature has some limitations, the approach offers the advantage of providing fine, sub-kilometer spatial scales, as conventional numerical weather models are tied to coarser grid sizes.

Recent weather forecast models have been adapted to spatial-temporal scales fine enough

Table 2.3.1. Major forest fire databases and inventories.

	Frequency	Coverage	Resolution	Reference	Name/organization
GFED	Monthly	1995–now	0.25° × 0.25°	Randerson *et al.*, 2012; Giglio *et al.*, 2013. http://www.globalfiredata.org	Global Fire Emissions Database (UMD, NASA)
GFAS	Daily	2001–now	0.5° × 0.5°	Kaiser *et al.*, 2012. https://apps.ecmwf.int/datasets/data/cams-gfas/	Global Fire Assimilation System (ECMWF-MACC)
FINN	Daily	2001–now	1 km × 1 km	Wiedinmyer *et al.*, 2011. https://www.acom.ucar.edu/Data/fire/	Fire Inventory from NCAR
FEER	Daily	2003–now	1° × 1°	Ichoku and Ellison, 2014. https://feer.gsfc.nasa.gov	Fire Energetics and Emissions Research (NASA)
QFED	Daily	Now (use in GOCART)	0.1° × 0.1°	Darmenov and Da Silva, 2015. https://gmao.gsfc.nasa.gov/research/science_snapshots/global_fire_emissions.php	Quick Fire Emissions Dataset (NASA)
GICC	Decadal–monthly	1900–2005	1° × 1°	Mieville *et al.*, 2010. http://accent.aero.jussieu.fr/GICC_metadata.php	Global Inventory for Chemistry-Climate Studies (ACCENT)

ACCENT, Atmospheric Composition Change, the European Network of Excellence; ECMWF, European Centre for Medium-Range Weather Forecasts; FEER, Fire Energetics and Emissions Research – NASA; FINN, Fire INventory from NCAR; GFAS, Global Fire Assimilation System; GFED, Global Fire Emissions Database; GICC, Global Inventory for Chemistry-Climate; MACC, Monitoring Atmospheric Composition and Climate; NCAR, National Center for Atmospheric Research; UMD, University of Maryland – Department of Atmospheric and Oceanic Science.
All websites accessed 2 January 2022.

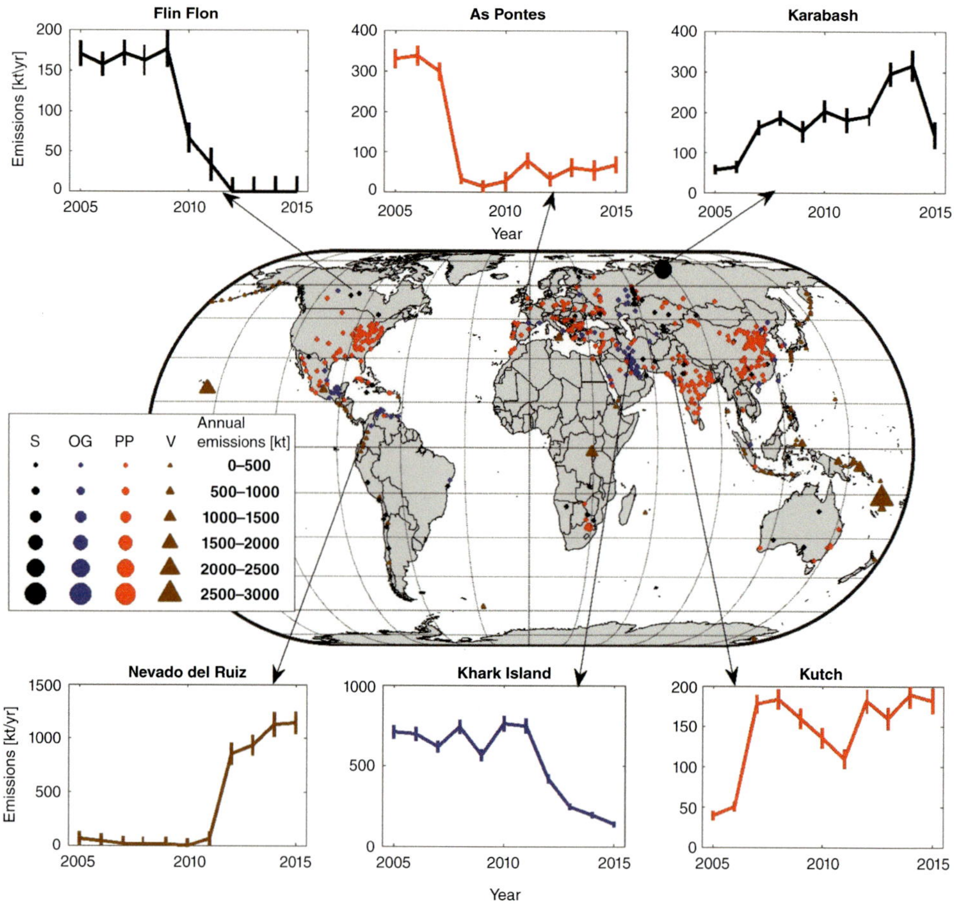

Fig. 2.3.1. Annual SO_2 emissions from 2005 to 2015 derived from the Ozone Monitoring Instrument (OMI) for approximately 500 locations worldwide, according to four source types: smelters (S), oil and gas operations (OG), coal-fired power plants (PP), and volcanoes (V). Time series are shown for six different locations. (From: Fioletov *et al.*, 2016.)

to capture intra-urban thermal variabilities (Leroyer *et al.*, 2014). This required the development of appropriate urban surface parametrizations, such as the Town Energy Balance (TEB) scheme (Masson, 2000). The TEB scheme adds proficiency to intermediate scale weather forecast systems such as the High-Resolution Deterministic Prediction System of ECCC (Milbrandt *et al.*, 2018). The combination of urban schemes with land surface data assimilation systems can take further advantage of thermal imagery to drive land surface models coupled to atmospheric models, with expected gains in urban-scale weather forecast (Carrera *et al.*, 2015).

Challenges and questions

There are opportunities and emerging operational capabilities for assimilating air quality-related EO observations over large areas into public health-related services. The North American FireWork system currently covers large areas of Canada and the USA on an operational basis (Pavlovic *et al.*, 2016; Chen *et al.*, 2019). One of its tasks consists of incorporating more detailed spatio-temporal EO data into the existing forest fire smoke forecast systems (more details on FireWork are provided on page 45). Another challenge is the assessment of EO capabilities with regard to linking source areas with vulnerable populations

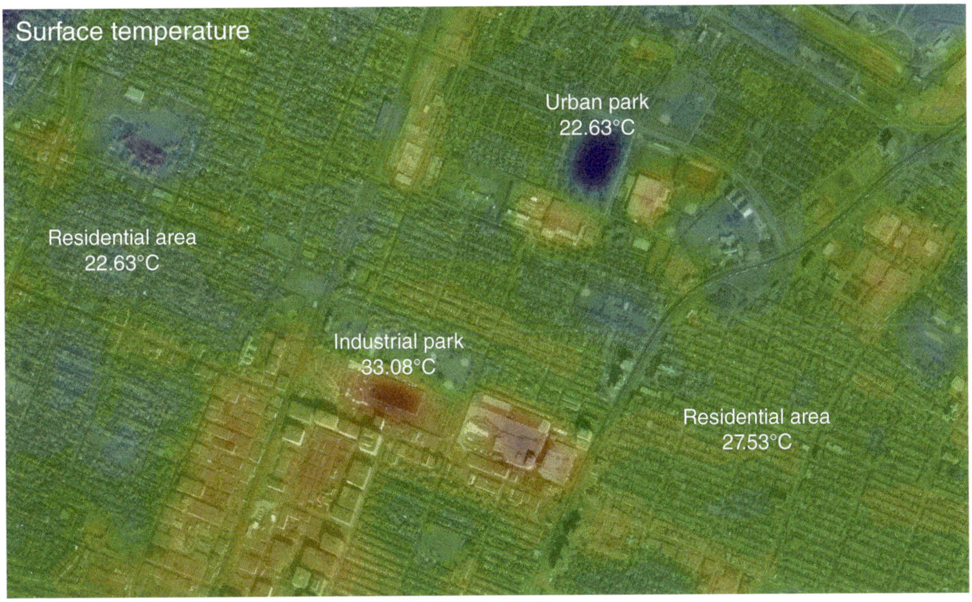

Fig. 2.3.2. Example of Landsat-8 Thermal Infrared Sensor (TIRS) thermal band measurements, dated 20 August 2013, superimposed on detailed satellite imagery of an urban area in Montreal, QC, showing the close relationship of land use categories and thermal conditions. Surface heat islands are depicted in red; cooler areas are represented in shades of green and blue. (From: Philippe Martin, ECCC.)

and outlining areas exposed to long-range transport of harmful particle plumes.

Challenges associated with the assessment and monitoring of UHIs are similar in nature in that relatively low-resolution thermal data need to be linked to detailed land use information as well as information on population distribution and density. Mapping, monitoring, and assessing actual urban land use and thermal conditions are tasks that can involve satellite-based measurements. The challenge is to determine spatio-temporal data requirements in conjunction with sensor system capabilities at a local scale.

Overall, for EO to support applied science related to public health concerns for air quality or heat, several questions and critical issues need to be addressed:

- What are the critical EO and geospatial data requirements for addressing large-area $PM_{2.5}$-related issues of air-borne diseases and chronic conditions, including remote human communities?
- What are the critical EO and geospatial data requirements for addressing urban area $PM_{2.5}$-related issues of air-borne diseases and chronic conditions, including densely populated areas?
- More specifically, what are the data definition and operational requirements that would optimize the value of real-time or near-real-time EO data for data assimilation into air quality forecast systems and urban-scale weather forecast systems?
- What EO data sources would lend themselves to detailed mapping of LST and UHIs, associated land use information, and population/built-up areas?
- How and at what level of spatial-temporal detail might these EO data sources and measurements be used for public health research?

Responses and options

Expert consultations yielded the following responses for potential options and actions regarding critical geospatial data requirements to address air-borne diseases and chronic conditions, with special reference to $PM_{2.5}$:

- Critical EO and GIS data for large-area $PM_{2.5}$ are aerosol optical depth (AOD) data from MODIS, GOES-16, and VIIRS, for example, and land use and land cover (surface permeability) products from Landsat.
- EO data sources for detailed LST and heat island maps include MODIS LST and Landsat thermal data, supplemented by airborne data.
- Spatial and temporal resolutions are challenging; AOD data are not useful in high albedo environments (e.g. snow-covered landscapes, desert areas); time of day of observation is an important factor.
- Model estimates of PM from AOD require validation using *in situ* $PM_{2.5}$ data, as AOD is not equivalent to PM in the boundary layer.
- Multiple thematic data sets are required, including: census data; meteorological data and models; vegetation index data; and socio-economic data (e.g. traffic density).
- Additional thematic data requirements for detailed urban $PM_{2.5}$ analysis include those related to tree canopies, urban morphology, hospital admissions, social media (using hashtags), population mobility (relative to location of exposure), and emission inventories.
- The usefulness of EO-based thermal measurements for public health research extends to short-term (daily) and long-term exposure monitoring capabilities, with the ability to tie spatio-temporal measurements to the most vulnerable areas and populations.
- The usefulness of EO and PM measurements could extend to measure diurnal variations and to improve national and provincial level emission inventories.

The American AirNow AQI for decision support and future EO missions

In the USA, the AQI offers daily reports on how clean or polluted the air is, and what associated health effects might be of concern. The AQI is calculated from five major air pollutants: ground-level ozone, particle pollution or PM, carbon monoxide, sulfur dioxide, and nitrogen dioxide. Since 1997, the AirNow data management and dissemination tool has provided up-to-date AQI information online for the USA and many parts of the world utilizing both ground and remotely sensed MODIS observations, notably on $PM_{2.5}$ exposure.

The AirNow software includes worldwide air quality mapping and multi-language capabilities that can be implemented in cities and regions around the globe. The system also allows decision makers to better communicate about air pollution, health, and sustainability goals, and to involve the public in efforts to improve air quality. MODIS sensor data help to fill many coverage gaps so that a large portion of the population and media outlets now have access to air quality information (Fig. 2.3.3). In the USA, federal, state, tribal, and local users, researchers, and air quality organizations provide data to and derive forecasts from the AirNow system. A group of international organizations and agencies provides overall direction, standardization, and scientific understanding for the collection, sharing, and disseminating of air quality data and forecasts and promoting AirNow development and growth in a self-supporting environment.

AirNow map products include $PM_{2.5}$ ground data and MODIS data algorithms and modeling required for $PM_{2.5}$ calibration. Time of day of observation is an important factor as there is a need to capture diurnal variations. The remote sensing data are not reliable for areas of high albedo, such as desert landscapes and snow-covered terrain. The AirNow system requires "big data" infrastructure for storage and processing, as well as skill and capacity among end users to utilize the final products knowledgably. It is a prime example of interagency cooperation to create a new product of significant societal health benefit.

There are several significant EO sensor developments underway in the USA, Europe, and Korea concerning pollution measurements. The TEMPO instrument represents the North American component of a satellite constellation with eventual global coverage provided by the Copernicus Sentinel-4 sensor system of the ESA and the Korean GEMS. TEMPO will make tropospheric pollution observations (O_3, NO_2, H_2CO) every daylight hour at high spatial resolution from geostationary orbit. The instrument was delivered in 2018 and the launch is planned for 2022. TEMPO data will be used to enhance pollution emission inventories, record population exposure, and assess effective emission-control strategies.[5]

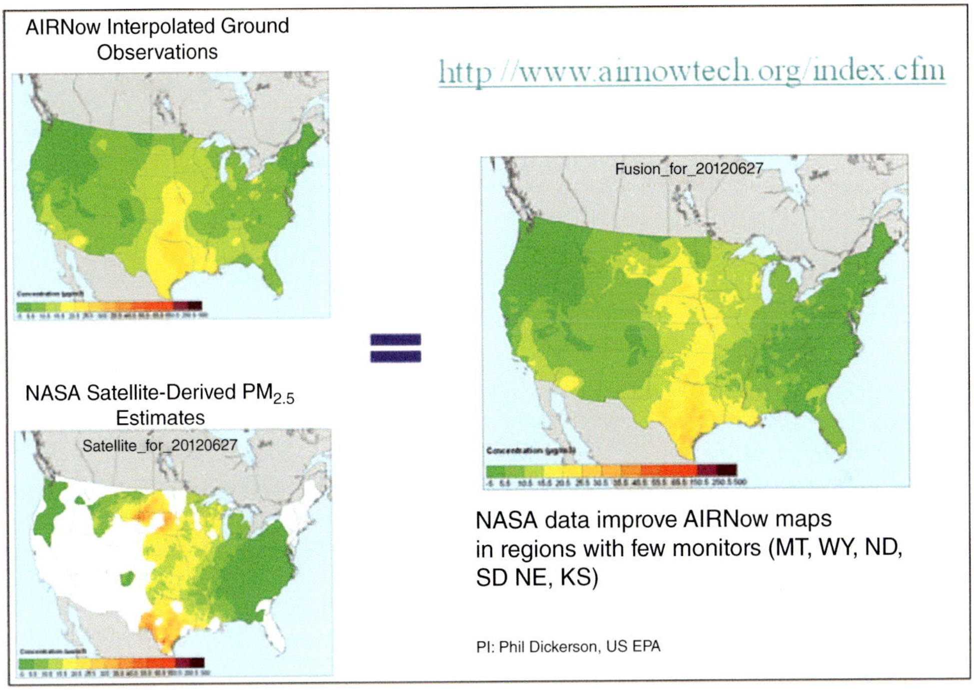

Fig. 2.3.3. The AirNow operational concept of using satellite data to augment missing $PM_{2.5}$ ground measurements in the USA. (From: https://www.airnowtech.org/index.cfm, accessed 3 January 2022.)

As part of its "Earth System Science Pathfinder" program, the NASA Jet Propulsion Laboratory is currently developing the future MAIA (Multi-Angle Imager for Aerosols) instrument whose primary objective is to assess links between different air-borne PM types and adverse birth outcomes, cardiovascular and respiratory disease, and premature deaths. Air-borne PM is a well-known health threat, but the relative toxicity of specific PM types is poorly understood. The sensor is a multi-angle spectro-polarimetric imaging instrument to be launched in 2022 for operation in a sun-synchronous Earth orbit to measure particle types, sizes, concentrations, and geolocation of atmospheric aerosols. The MAIA EO data generated from these radiometric and polarimetric measurements will be used to integrate air quality observations and geostatistical models, surface PM monitors, health records, and epidemiology to better understand the links between these pollutants and aerosols and human health. MAIA stakeholders include the US EPA, the CDC, and NOAA.

NOAA's recently launched Geostationary Operational Environmental Satellites GOES-16 and GOES-17 are both equipped with an Advanced Baseline Imager (ABI) for accurate quantification of AOD throughout the day over the USA and southern Canada. The increased number of spectral bands of ABI compared with previous GOES satellites is expected to significantly improve AOD measurements over land areas.

The Multi-Viewing Multi-Channel Multi-Polarization Imager (3MI) is the ESA's analogue to the American MAIA instrument. Both instruments are multi-angle imaging spectropolarimeters. However, 3MI has a much wider swath, a much longer mission lifetime of 7 years, and will consist of a series of three satellites that are expected to provide AOD measurements over a 20-year period from 2021 to 2042.

The Canadian FireWork system for air quality forecast related to wildfire emission of pollutants

Wildfire emissions are a growing concern for public safety in North America. Smoke plumes contain high concentrations of primary $PM_{2.5}$. Fire emissions also contribute to the formation of ground-level ozone and secondary organic aerosols. They can be the dominant contributors to adverse air quality issues during the summer, with long transport impacting other parts of the continent and beyond. Large-scale fires such as those in western Canadian provinces and in California in recent years can produce significantly more emissions than those from anthropogenic sources in the region. Furthermore, climate changes are expected to further increase the frequency, size, and duration of wildland fires across North America (Wotton *et al.*, 2017).

Given the sporadic nature of wildland fires and the complexity of quantifying emissions, monitoring transport, and assessing the chemical transformation of smoke, it is difficult to forecast the impact and severity on air quality. ECCC developed a numerical air quality forecast system as a numerical guidance tool. The FireWork smoke prediction system combines near-real-time EO data and the core GEM-MACH chemical transport model, in effect simulating the transport and atmospheric loading of $PM_{2.5}$ as a result of wildland fires.[6]

Expected outcomes and impacts

The FireWork system has been operational since May of 2016 and generates daily forecasts during the Canadian fire season from April to October. Twice a day (00z and 12z UTC), the system produces a North America-wide 48-h model forecast of air pollution as $PM_{2.5}$ surface and total column concentrations. In Canada, the simulation results provide numerical guidance on regional air quality conditions to forecasters in regional offices of the Meteorological Service of Canada (MSC). They issue public air quality forecast in the form of the AQHI (Stieb *et al.*, 2008), as well as smoke-related special weather bulletins.

FireWork is a collaborative product by Natural Resources Canada (Chen *et al.*, 2019). Near-real-time information on fire locations and estimated fuel consumptions across North America are obtained from the Canadian Wildland Fire Information System (CWFIS), which is operated by the Canadian Forest Service of Natural Resources Canada (Lee *et al.*, 2002). Fire locations, or hotspots, are retrieved from EO data originating from NASA's MODIS and VIIRS satellite sensors or NOAA's AVHRR. The near-real-time EO data are processed with the Canadian Fire Emissions Prediction System (CFFEPS) and with hourly forecast weather conditions to estimate fire behavior, fuel consumption, and fire emissions at hotspot locations for the upcoming 72 h. Heat energy from hotspots is then estimated to adjust fire plume injection heights that are critical in dispersion transport modeling. And finally, hourly fire emissions are incorporated together with anthropogenic emissions inventory for GEM-MACH chemical transport model simulation.

An excess of pollution forecasted by FireWork is considered to be the result of direct fire emission. This excess is calculated by comparing with the MSC's Regional Air Quality Deterministic Prediction System without biomass burning emissions. This simple – but computationally expensive – approach allows individual wildfire smoke plumes to be isolated, followed, and forecast.

Behind the scenes, the FireWork system provides guidance to ECCC meteorologists issuing air quality forecasts and warnings. In addition, key operational model products from FireWork are available to the public as part of Canada's weather information website.[7] Products include daily average or maximum forecast maps. Hourly animations are available to show the modeled trajectory of smoke plume and concentration variability from identified fire activity across the model domain (Fig. 2.3.4).

Near-real-time EO data constitute an essential element of the FireWork system due to the large spatial extent of the model domain across the North American continent. EO data are required for the identification of the sporadic wildfire locations and the approximate timing and size of the fire. This information is critical for quantifying fire emissions for model input. In addition, EO data provide timely updates that are often not available through surface observations, as well as a higher degree of data uniformity that is often preferable to the varying methodologies and timeliness of fire activity records from different provincial or state government entities. This combination of EO and model data processing was demonstrated in a recent study that quantified carbon monoxide,

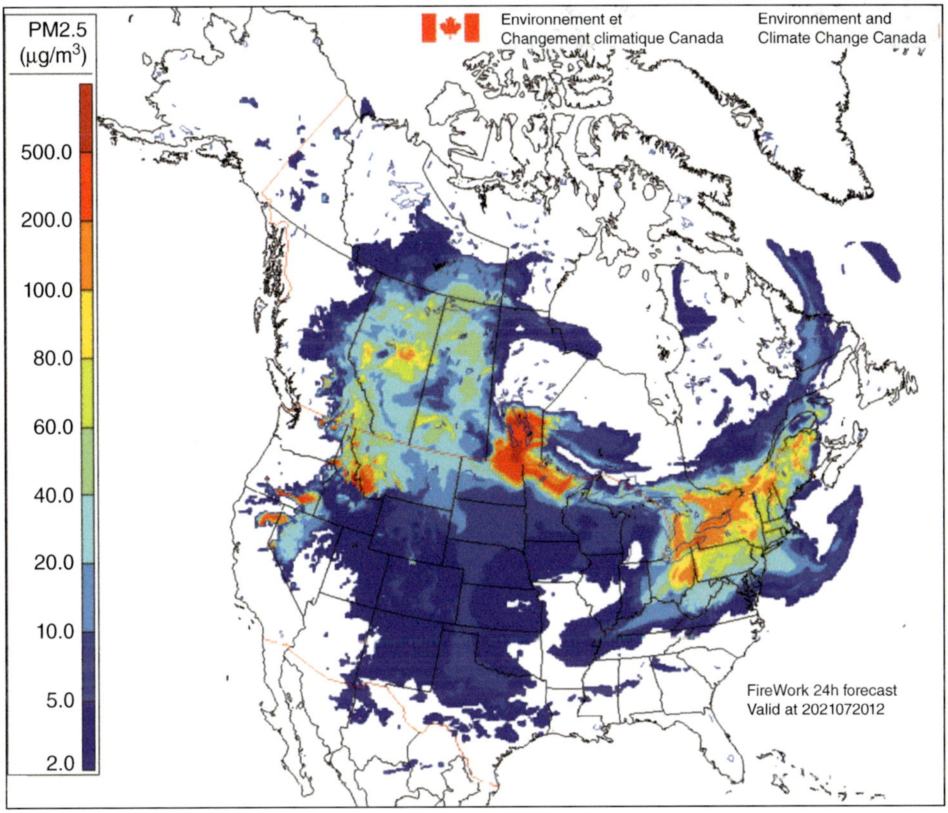

Fig. 2.3.4. Example of a FireWork smoke forecast product, showing the 24-h forecast of the contribution of wildland fires to surface $PM_{2.5}$ concentrations, valid at 12:00 UTC on 20 July 2021.

ammonia, and nitrogen dioxide emissions from the 2016 Horse River wildfire in the Fort Mc-Murray area (Adams *et al.*, 2019). EO data are also an important source of information for verification of modeling parameterization such as plume injection height, which is an important parameter in atmospheric dispersion modeling (Griffin *et al.*, 2020).

The FireWork system has been demonstrated to be a valuable tool in providing guidance for regional air quality during smoke events from wildland fires across North America (McLean *et al.*, 2015; Yuchi *et al.*, 2016). Regional forecasters, emergency first responders, and decision makers use FireWork results as guidance for assessing evacuation orders and issuing special health bulletins. The public uses FireWork to visualize the long-range transport of wildland fire smoke plumes. In addition, the FireWork archive is utilized by scientists in Canada to study the potential exposure and associated health impact of atmospheric pollutants from biomass burning activities (Munoz-Alpizar *et al.*, 2017).

Technical considerations and perspectives for system operation

Technical considerations of implementing and operating the FireWork system include the selection of EO data, adoption of prior technical programs, as well as computational requirements.

The EO data needed for determining active fire locations, or hotspots, originate from retrievals from NASA MODIS and VIIRS, and NOAA AVHRR, with support from the University of Maryland and the US Forest Service Remote Sensing Application Center. While the FireWork system uses these fire detection products combined with a "bottom-up" approach to fuel consumption based on a land-vegetation model, similar systems have used a "top-down" approach that relies on correlating remotely sensed FRP to fire energy and fuel consumption (Mota and Wooster, 2018). As the science evolves, forecasting systems will adjust to the most appropriate scheme for their operations.

The development of the Canadian FireWork system is an extension of existing programs in two federal departments, ECCC and Natural Resources Canada. As a comprehensive numerical weather prediction system, FireWork requires a high-performance computing (HPC) environment with critical 24/7 support for its operation. In addition, the system relies on the operation of the fire behavior prediction system run by Natural Resources Canada's CWFIS. System improvements require dedicated resources for model development, EO data processing, and model forecast operations. Ultimately, the rationale for running and utilizing the FireWork system lies in its contribution to a comprehensive Canadian air quality forecast service that is delivered by ECCC. Its resources are dedicated to developing appropriate products for both the public and partners.

Effective observation by air-borne and satellite-based optical sensors may at times be obscured by cloud and smoke, thus reducing the usefulness of the product when the emissions from active fire locations are not registered. Improving spatial resolution would be useful because it could allow for assigning fire detection to land fuel with higher accuracy.

Currently there is no distinction in retrievals on fire activity between wildland forest fires, prescribed fires, and agriculture fires. These types are usually assigned post-retrieval based on input fuels maps, which can differ greatly across political boundaries. The distinction between fire types can be important for process-based emission estimates (Liu *et al.*, 2017).

The timeliness of the EO data is also an important factor because the emissions data introduced into the forecast model should be as recent as possible. Using the current FireWork regime, several hours can pass before a detectable fire is captured by an EO sensor and related data are available to the forecast.

Finally, a fair amount of effort is currently invested in screening recurrent false detections. False detections are likely to increase as the spatial resolution of the satellite sensor increases. A comprehensive approach to solving this problem would be highly beneficial.

From a technical point-of-view, there are opportunities to improve the FireWork products by way of including additional EO data and further R&D activities in the validation and modeling phases. There are EO sensors and products, e.g. the OMI[8] aboard the Aura spacecraft and the European TROPOMI[9] aboard the Sentinel-5 spacecraft, that can detect fire-related pollutant concentrations of CO, NO_2, and HCHO. These can be used to infer fire emissions as an alternative to the fire radiative power-based approach.

And finally, high-resolution EO products can help address uncertainty in active fire plume-rise height estimates (Paugam *et al.*, 2016).

Findings, opportunities for collaboration, and conclusion

This section examines several aspects of the use of EO for applications related to air quality and weather extremes as they relate to public health issues. Air quality is a global public health issue. According to the WHO, 9 out of 10 people breathe air containing high levels of pollutants; it is estimated that around 7 million people die every year from exposure to fine particles in polluted air, which can lead to stroke, heart disease, lung cancer, chronic obstructive pulmonary diseases, and respiratory infections, including pneumonia (WHO, 2018). Tools are needed to better assess and quantify the risks associated with poor air quality. These tools must be available for health administrations and be accessible to the public. It would be desirable to integrate existing and future systems with public health systems, tools, or services such as surveillance, emergency management, and prevention and response actions. It would also be an asset to better communicate the level of risk to medical staff in hospitals and clinics concerned with the prevention, diagnosis, and treatment of patients.

The science is evolving from using EO to generate static or time-averaged information about surface pollutant concentrations or thermal anomalies, to integrating EO into more dynamic systems that support the generation of real-time forecast products. This development is illustrated by the emergence of air quality data assimilation capabilities for air quality forecast models, and land surface data assimilation systems that are coupled with urban-scale weather forecast models. This is also true regarding the growing demand for real-time air quality mapping data and the new requirement to systematically characterize air pollutant emissions from wildfires for applications in air quality forecast and reporting.

Current priorities in EO missions provide an appropriate first response to these needs, with upcoming missions adding valuable measurements of tropospheric pollutant concentrations at a finer spatial resolution (e.g. TROPOMI) and at a finer temporal resolution (e.g. TEMPO). These advances are complemented by increased access to AOD data (e.g. MODIS over the recent GOES-16 and GOES-17 satellites). Experts around the world are collaborating to advance the science for the valuation of these EO data sets into air quality data assimilation systems, with the ultimate goal of optimizing the representation of air quality at any time and any location, maximizing the benefits for air quality forecast and advancing air pollutant exposure studies.

The detection and characterization of wildland fires would largely benefit from better spatially and temporally resolved remote sensing. New instruments aboard geostationary satellites, such as GOES, can deliver high-frequency measurements, which are extremely useful for rate of spread calculations. However, they are limited in spatial resolution and only provide coverage over subpolar latitudes. One alternative would be a constellation of polar-orbiting satellite instruments. This approach was selected by the Canadian Space Agency, Natural Resources Canada, and ECCC, who worked together to develop the WildFireSat mission proposal (Van Mierlo, 2019). Commercial initiatives have also been put forward, such as the FireSat constellation,[10] which has proposed paid access to up to 200 polar-orbiting thermal sensors.

As air quality is a global issue, operators of relevant EO platforms are encouraged to collaborate and facilitate access to their air quality-related EO data. This aligns with the principles of data sharing stated in the Convention of the WMO and the mission of GEO. The WMO also offers a framework to develop and implement appropriate standards for sharing air quality data that would optimize its use across meteorological centers.

Over the coming decade, air quality and heat will continue to be a very dynamic area of research and science applications. It will rely heavily on EO data to successfully address the need to deliver timely and reliable information to decision makers and public health authorities.

Notes

1 https://www.canada.ca/en/environment-climate-change/services/air-pollution/monitoring-networks-data/canadian-air-precipitation.html and https://data.ec.gc.ca/data/air/monitor/national-air-pollution-surveillance-naps-program/ (accessed 19 January 2022).

2 https://directory.eoportal.org/web/eoportal/satellite-missions/t/tempo (accessed 3 January 2022).

3 https://ephtracking.cdc.gov/showUVTracking (accessed 3 January 2022).

4 https://airquality.gsfc.nasa.gov/ (accessed 3 January 2022).

5 http://tempo.si.edu/overview.html (accessed 3 January 2022).

6 https://weather.gc.ca/firework/ (accessed 3 January 2022).

7 https://weather.gc.ca/firework (accessed 3 January 2022).

8 https://aura.gsfc.nasa.gov/omi.html (accessed 3 January 2022).

9 http://www.tropomi.eu/ (accessed 3 January 2022).

10 FireSat – Global Fire Monitoring Center: http://gfmc.online/current/FIRESAT-Brochure-2017.pdf (accessed 3 January 2022).

References

Adams, C., McLinden, C.A., Shephard, M.W., Dickson, N., Dammers *et al.* (2019) Satellite-derived emissions of carbon monoxide, ammonia, and nitrogen dioxide from the 2016 Horse River wildfire in the Fort McMurray area. *Atmospheric Chemistry and Physics* 19, 2577–2599. DOI: https://doi.org/10.5194/acp-19-2577-2019.

Carrera, M.L., Bélair, S., and Bilodeau, B., (2015) The Canadian Land Data Assimilation System (CaLDAS): Description and Synthetic Evaluation Study. *Journal of Hydrometeorology* 16, 1293–1314. DOI: https://doi.org/10.1175/JHM-D-14-0089.1.

Chakraborty, T. and Lee, X. (2019) A simplified urban-extent algorithm to characterize surface urban heat islands on a global scale and examine vegetation control on their spatiotemporal variability. *International Journal of Applied Earth Observation and Geoinformation* 74, 269–280. DOI: https://doi.org/10.1016/j.jag.2018.09.015.

Chance, K., Liu, X., Suleiman, R.M., Flittner, D.E., Al-Saadi, J. *et al.* (2013) Tropospheric emissions: Monitoring of pollution (TEMPO), Proc. SPIE 8866, Earth Observing Systems XVIII. DOI: https://doi.org/10.1117/12.2024479.

Chen, J., Anderson, K., Pavlovic, R., Moran, M.D., Englefield, P. *et al.* (2019) The FireWork v2.0 air quality forecast system with biomass burning emissions from the Canadian Forest Fire Emissions Prediction System v2.03. *Geoscientific Model Development* 12, 3283–3310. DOI: https://doi.org/10.5194/gmd-12-3283-2019.

Darmenov, A.S. and Da Silva, A.M. (2015) The Quick Fire Emissions Dataset (QFED): Documentation of versions 2.1, 2.2 and 2.4. *NASA Technical Report Series on Global Modeling and Data Assimilation* 38. Available at: https://ntrs.nasa.gov/api/citations/20180005253/downloads/20180005253.pdf (accessed 5 August 2021).

De Sario, M., Katsouyanni, K., and Michelozzi, P. (2013) Climate change, extreme weather events, air pollution and respiratory health in Europe. *European Respiratory Journal* 42, 826–843. https://doi.org/10.1183/09031936.00074712.

Duncan, B. Prados A.I., Lamsal L.N., Liu Y., Streets D.G. *et al.* (2014) Satellite data of atmospheric pollution for U.S. air quality applications: Examples of applications, summary of data end-user resources, answers to FAQs, and common mistakes to avoid. *Atmospheric Environment* 94, 647–662. https://doi.org/10.1016/j.atmosenv.2014.05.061.

Fioletov, V.E., McLinden, C.A., Krotkov, N., Li, C., Joiner, J. *et al.* (2016) A global catalogue of large SO_2 sources and emissions derived from the Ozone Monitoring Instrument. *Atmospheric Chemistry and Physics* 16, 11497–11519. DOI: https://doi.org/10.5194/acp-16-11497-2016.

Giglio, L., Randerson, J.T., and van der Werf, G.R. (2013), Analysis of daily, monthly, and annual burned area using the fourth-generation global fire emissions database (GFED4). *Journal of Geophysical Research: Biogeosciences* 118, 317–328. DOI: https://doi.org/10.1002/jgrg.20042.

Giguère, M. (2009) *Literature Review of Urban Heat Island Mitigation Strategies*. Institut National de Santé Publique du Québec. Available at: https://www.inspq.qc.ca/sites/default/files/publications/1513_urbanheatislandmitigationstrategies.pdf (accessed 3 August 2021).

Griffin, D., Sioris, C., Chen, J., Dickson, N., Kovachik, A. *et al.* (2020) The 2018 fire season in North America as seen by TROPOMI: Aerosol layer height validation and evaluation of model-derived plume heights. *Atmospheric Measurement Techniques* 13. DOI: https://doi.org/10.5194/amt-2019-411.

Ichoku, C. and Ellison, L. (2014) Global top-down smoke-aerosol emissions estimation using satellite fire radiative power measurements. *Atmospheric Chemistry and Physics* 14, 6643–6667. DOI: https://doi.org/10.5194/acp-14-6643-2014.

Kaiser, J.W., Heil, A., Andreae, M.O., Benedetti, A., Chubarova, N. *et al.* (2012) Biomass burning emissions estimated with a global fire assimilation system based on observed fire radiative power. *Biogeosciences* 9, 527–554. DOI: https://doi.org/10.5194/bg-9-527-2012.

Le, T.H., Thanh Nguyen, T.N., Lasko, K., Ilavajhala, S., Vadrevu, K.P. *et al.* (2014) Vegetation fires and air pollution in Vietnam. *Environmental Pollution* 195, 267–275. DOI: https://doi.org/10.1016/j.envpol.2014.07.023.

Lee, B., Alexander, M., Hawkes, B., Lynham, T., Stocks, B. *et al.* (2002) Information systems in support of wildland fire management decision making in Canada. *Computers and Electronics in Agriculture* 37, 185–198. DOI: https://doi.org/10.1016/S0168-1699(02)00120-5.

Leroyer, S., Bélair, S., Husain, S.Z., and Mailhot, J. (2014) Subkilometer numerical weather prediction in an urban coastal area: A case study over the Vancouver Metropolitan Area. *Journal of Applied Meteorology and Climatology* 53, 1433–1453. DOI: https://doi.org/10.1175/JAMC-D-13-0202.1.

Levelt, P.F., Joiner, J., Tamminen, J., Veefkind, J.P., Bhartia, P.K. *et al.* (2018) The Ozone Monitoring Instrument: Overview of 14 years in space. *Atmospheric Chemistry and Physics* 18, 5699–5745. DOI: https://doi.org/10.5194/acp-18-5699-2018.

Liu, X., Huey, L.G., Yokelson, R.J., Selimovic, V., Simpson, I.J. *et al.* (2017) Airborne measurements of western U.S. wildfire emissions: Comparison with prescribed burning and air quality implications.

Journal of Geophysical Research Atmospheres 122, 6108–6129. DOI: https://doi.org/10.1002/2016JD026315.

Liu, F., Choi, S., Li, C., Fioletov, V.E., McLinden, C.A., *et al.* (2018) A new global anthropogenic SO_2 emission inventory for the last decade: A mosaic of satellite-derived and bottom-up emissions, *Atmospheric Chemistry and Physics: Discussions* 18. DOI: https://doi.org/10.5194/acp-2018-331.

Martin, P., Baudouin, Y., and Gachon, P. (2015) An alternative method to characterize the surface urban heat island. *International Journal of Biometeorology* 59, 849–861. DOI: https://doi.org/10.1007/s00484-014-0902-9.

Masson, V. (2000) A physically-based scheme for the urban energy budget in atmospheric models. *Boundary-Layer Meteorology* 94, 357–397. https://doi.org/10.1023/A:1002463829265.

McLean, K., Yao, J., and Henderson, S. (2015) An evaluation of the British Columbia Asthma Monitoring System (BCAMS) and $PM_{2.5}$ Exposure Metrics during the 2014 Forest Fire Season. *International Journal of Environmental Research and Public Health* 12, 6710–6724. DOI: https://doi.org/10.3390/ijerph120606710.

McLinden, C.A., Fioletov, V., Shephard, M.W., Krotkov, N., Li, C. *et al.* (2016) Space-based detection of missing sulfur dioxide sources of global air pollution. *Nature Geoscience* 9, 496–500. DOI: https://doi.org/10.1038/ngeo2724.

Mieville, A., Granier, C., Liousse, C., Guillaume, B., Mouillot, F. *et al.* (2010) Emissions of gases and particles from biomass burning during the 20th century using satellite data and an historical reconstruction. *Atmospheric Environment* 44. 1469–1477. DOI: http://dx.doi.org/10.1016/j.atmosenv.2010.01.011.

Milbrandt, J., Leroyer, S., Paquin-Ricard, D., Faucher, M., and Zhang, S. (2018) High Resolution Deterministic Prediction System (HRDPS). *Canadian Meteorological Centre Technical Note.* Available at: http://collaboration.cmc.ec.gc.ca/cmc/CMOI/product_guide/docs/lib/technote_hrdps-500_20180918_e.pdf (accessed 15 June 2021).

Mota, B. and Wooster, M.J. (2018) A new top-down approach for directly estimating biomass burning emissions and fuel consumption rates and totals from geostationary satellite fire radiative power (FRP). *Remote Sensing of Environment* 206, 45–62. DOI: https://doi.org/10.1016/j.rse.2017.12.016.

Munoz-Alpizar, R., Pavlovic, R., Moran, M., Chen, J., Gravel, S. *et al.* (2017) Multi-year (2013–2016) $PM_{2.5}$ wildfire pollution exposure over North America as determined from operational air quality forecasts. *Atmosphere* 8, 1–24. DOI: http://dx.doi.org/10.3390/atmos8090179.

Organisation for Economic Co-operation and Development (OECD) (2014) *The Cost of Air Pollution: Health Impacts of Road Transport. OECD, Paris.* DOI: http://dx.doi.org/10.1787/9789264210448-en.

Paugam, R., Wooster, M., Freitas, S., and Val Martin, M. (2016) A review of approaches to estimate wildfire plume injection height within large-scale atmospheric chemical transport models. *Atmospheric Chemistry Physics* 16, 907–925. DOI: https://doi.org/10.5194/acp-16-907-2016.

Pavlovic, R., Chen, J., Anderson, K., Moran, M.D., Beaulieu, P.-A. *et al.* (2016) The FireWork air quality forecast system with near-real-time biomass burning emissions: Recent developments and evaluation of performance for the 2015 North American wildfire season. *Journal of the Air & Waste Management Association* 66, 819–841. DOI: https://doi.org/10.1080/10962247.2016.1158214.

Peng, Z., Liu, Z., Chen, D., and Ban, J. (2017) Improving $PM_{2.5}$ forecast over China by the joint adjustment of initial conditions and source emissions with an ensemble Kalman filter. *Atmospheric Chemistry and Physics* 17, 4837–4855. DOI: https://doi.org/10.5194/acp-17-4837-2017.

Randerson, J.T., Chen, Y., van der Werf, B.M., Rogers, D., and Morton C. (2012) Global burned area and biomass burning emissions from small fires. *Journal of Geophysical Research: Biogeosciences* 117. DOI: https://doi.org/10.1029/2012JG002128.

Rochon, Y.J., Sitwell, M., and Cho, Y-M. (2019) A study on harmonizing total ozone assimilation with multiple sensors. *Atmospheric Chemistry and Physics* 19, 9431–9451. DOI: https://doi.org/10.5194/acp-19-9431-2019.

Schroeder, W., Oliva, P., Giglio, L., Quayle, B., Lorenz, E. *et al.* (2016) Active fire detection using Landsat-8/OLI data. *Remote Sensing of Environment* 185, 210–220. DOI: https://doi.org/10.1016/j.rse.2015.08.032.

Shen, L., Guo, X., and Xiao, K. (2014) Spatiotemporally characterizing urban temperatures based on remote sensing and GIS analysis: A case study in the city of Saskatoon (SK, Canada). *Open Geosciences* 7. DOI: https://doi.org/10.1515/geo-2015-0005.

Sofowote, U. and Dempsey, F. (2015) Impacts of forest fires on ambient near-real-time $PM_{2.5}$ in Ontario, Canada: Meteorological analyses and source apportionment of the July 2011–2013 episodes. *Atmospheric Pollution Research* 6, 1–10. DOI: https://doi.org/10.5094/APR.2015.001.

Stieb, D.M., Burnett, R.T., Smith-Doiron, M., Brion, O., Shin, H.H. *et al.* (2008) A new multipollutant, no-threshold air quality health index based on short-term associations observed in daily time-series analyses. *Journal of the Air & Waste Management Association* 58, 435–450. DOI: https://doi.org/10.3155/1047-3289.58.3.435.

Streets, D.G., Canty, T., Carmichael, G.R., de Foy, B., Dickerson, R.R. *et al.* (2013) Emissions estimation from satellite retrievals: A review of current capability. *Atmospheric Environment* 77, 1011–1042. DOI: https://doi.org/10.1016/j.atmosenv.2013.05.051.

Tereszchuk, K.A., Rochon, Y.J., McLinden, C.A., and Vaillancourt, P.A. (2018) Optimizing UV Index determination from broadband irradiances. *Geoscientific Model Development* 11, 1093–1113. DOI: https://doi.org/10.5194/gmd-11-1093-2018.

Van Mierlo, H. (2019) Unlocking the potential for a global wildfire monitoring constellation. https://cnrfire2019.eu/wp-content/uploads/2019/10/Van-Mierlo_Earsel2019.pdf (accessed 16 June 2021).

World Health Organization (WHO) (2018) Ambient (outdoor) air pollution fact sheet. Available at: https://www.who.int/news-room/fact-sheets/detail/ambient-(outdoor)-air-quality-and-health (accessed 3 August 2021).

Wiedinmyer, C., Akagi, S.K., Yokelson, R.J., Emmons, L.K., Al-Saadi, J.A. *et al.* (2011) The Fire INventory from NCAR (FINN): A high resolution global model to estimate the emissions from open burning. *Geoscientific Model Development* 4, 625–641. DOI: https://doi.org/10.5194/gmd-4-625-2011.

Wotton, B.M., Flannigan, M.D., and Marshall, G.A. (2017) Potential climate change impacts on fire intensity and key wildfire suppression thresholds in Canada. *Environmental Research Letters* 12. DOI: https://doi.org/10.1088/1748-9326/aa7e6e.

Yuchi, W., Yao, J., McLean, K.E., Stull, R., Pavlovic, R. *et al.* (2016) Blending forest fire smoke forecasts with observed data can improve their utility for public health applications. *Atmospheric Environment* 145, 308–317. DOI: https://doi.org/10.1016/J.ATMOSENV.2016.09.049.

Zoogman, P., Liu, X, Suleiman R.M., Pennington, W.F., Flittner, D.E. *et al.* (2017) Tropospheric emissions: Monitoring of pollution (TEMPO). *Journal of Quantitative Spectroscopy and Radiative Transfer* 186 17–39. DOI: https://doi.org/10.1016/j.jqsrt.2016.05.008.

2.4 Water-borne Diseases: EO System for the Coastal Monitoring of Non-Cholera *Vibrios*

Jan C. Semenza[1], Joaquin Trinanes[2,3,4], Stéphanie Brazeau[5], Philippe Berthiaume[5], Marie-Josée Champagne[5], Joacim Rocklöv[6], and Jaime Martinez-Urtaza[7]

[1]Scientific Assessment, SRS, European Centre for Disease Prevention and Control (ECDC), Solna, Sweden; [2]University of Santiago de Compostela, CRETUS Institute, Santiago de Compostela, Spain; [3]University of Miami, Cooperative Institute for Marine and Atmospheric Studies, Miami, Florida, USA; [4]National Oceanic and Atmospheric Administration, Atlantic Oceanographic and Meteorological Laboratory, Miami, Florida, USA; [5]Public Health Risk Sciences Division, National Microbiology Laboratory, Public Health Agency of Canada, Saint-Hyacinthe, Québec, Canada; [6]Umeå University, Umeå, Sweden; [7]Department of Genetics and Microbiology, Universitat Autònoma de Barcelona, Barcelona, Spain

Context, state of knowledge, challenges, and responses

Infections caused by *Vibrio* species are a major concern for public health worldwide. It has been estimated that *Vibrio cholerae* infections cause 2.9 million cases annually and 95,000 deaths in developing countries (Ali *et al.*, 2015). In industrialized nations, *V. cholerae* is much less of a concern, but infections caused by non-cholera vibrios (NCV), such as *Vibrio parahaemolyticus* or *Vibrio vulnificus*, have increased and are expected to continue increasing due to climate change (Baker-Austin *et al.*, 2013; Schijven *et al.*, 2013; Semenza *et al.*, 2012a, 2012b, 2017; Taylor *et al.*, 2018).

Although a few European countries and the US CDC conduct NCV surveillance, there is a pressing need for improved global surveillance data on NCV infections and its presence and concentration in the environment over time and space.

NCVs are widely distributed in aquatic environments, and environmental factors play an important role in their growth (Johnson, 2015). Because global climate changes are affecting variables such as sea surface temperature (SST) and sea surface salinity (SSS), areas previously unsuitable for the survival and growth of NCV are opening up – making timely data acquisition even more critical (Baker-Austin *et al.*, 2010, 2013, 2017; Semenza *et al.*, 2012). Unfortunately, data acquisition using *in situ* sampling is both expensive and time-consuming, which is why EO technologies have become so useful in measuring environmental parameters that influence the growth of NCV (Grimes *et al.*, 2014). The combination of remote sensing data and models could provide an early warning system for human health risks associated with NCV (Baker-Austin *et al.*, 2013; Konrad *et al.*, 2017).

This section presents the benefits and challenges of using remote sensing data to predict human health risks associated with NCV, and it discusses the data and technological requirements for generating these data using the ECDC Vibrio Map Viewer developed by the ECDC as an example (Semenza *et al.*, 2017).

NCVs, environmental factors, and satellite-based monitoring capabilities

NCV species can cause gastroenteritis through the consumption of contaminated seafood or the infection of wounds due to direct exposure to the NCV present in marine water (Semenza *et al.*, 2012a). These wound infections are potentially serious and can result in septicemia and death. Among the 10 species of NCV capable of causing diseases in humans, *V. parahaemolyticus* (Vp), *V. alginolyticus*, and *V. vulnificus* (Vv) are the greatest concern given the number of human cases reported and/or the relatively high rate of fatality (Janda *et al.*, 1988; CDC, 2014; Government of Canada, 2015).

NCVs have worldwide distribution (Ceccarelli *et al.*, 2013; Kokashvili *et al.*, 2015). They are most frequently found either as free-living aquatic bacteria or associated with various aquatic substrates in brackish marine environments (Sarkar *et al.*, 1985; Venkateswaran *et al.*, 1989; McCarter, 1999; Gonzalez *et al.*, 2014). The concentration of NCV in sea water is an important factor driving the human risk of infection from both direct water contact and from seafood. Shellfish are often a source of infection because many species obtain nutrients by filtering water, which can accumulate and concentrate NCV bacteria in the shellfish (Hlady and Klontz, 1996).

Although the ecology of different NCV species is still under investigation, there is a general consensus that an association between NCV concentrations and water temperature exists (Kaneko and Colwell, 1973; Williams and Larock 1985; DePaola *et al.*, 2003; Pfeffer *et al.*, 2003; Thompson *et al.*, 2004; Semenza *et al.*, 2012a, 2012b; Siboni *et al.*, 2016). More specifically, it has been observed that once a temperature threshold for growth is reached, there is a positive correlation between increasing SST and NCV concentrations (Semenza *et al.*, 2012a, 2012b). Several authors have reported that NCV species are most abundant in the warmest periods of the year (Kelly and Stroh, 1988; DePaola *et al.*, 2003; Nigro *et al.*, 2011; Urquhart *et al.*, 2016). This seasonality in NCV concentration has been observed in shellfish, water, and sediments (Johnson *et al.*, 2010). Kaneko and Colwell (1973) found a threshold temperature of 14°C, above which Vp is activated in the environment, and Kelly and Stroh (1988) found little evidence for the presence of Vp in the environment below 17°C. In the colder months, NCV can still be isolated, albeit in much smaller numbers, from sediments, shellfish, and other aquatic organisms, and a viable but non-culturable form has been described for many NCV species (Oliver *et al.*, 1995; Thompson *et al.*, 2004; Amel *et al.*, 2008; Coutard *et al.*, 2007; Zhong *et al.*, 2009).

Salinity is an environmental factor identified as a determinant of the presence of NCV (Semenza *et al.*, 2012a, 2012b). NCV can be found in estuarine and coastal waters with low to moderate salinity (Janda *et al.*, 1988), and differences in salinity affect the presence of NCV. In one study, 64% of the observed variation in Vp concentration in water and 76% in oysters was attributed to salinity differences (Zimmerman *et al.*, 2007). In another study, an upsurge in NCV densities was observed with increasing salinity up to an optimum salinity point (e.g. for Vv: 27 parts per trillion [ppt] NaCl), above which a decrease was observed with increased salinity (Johnson *et al.*, 2010). In yet another study, salinity was found to explain an additional 10% in the variability in Vv once temperature was already accounted for (Motes *et al.*, 1998). In contrast, some authors did not find a significant impact of salt concentration either in oysters or in water (DePaola *et al.*, 2003; Nigro *et al.*, 2011). Nevertheless, low to moderate salinity is necessary for the presence of NCV and, as such, delineates the geographic distribution of NCV in marine environments.

The association between NCV and other environmental factors or indicators such as turbidity, plankton, chlorophyll, nutrients, and oxygen have been studied but the relationship was weaker than with temperature and salinity (Zimmerman *et al.*, 2007; Johnson *et al.*, 2010; Nigro *et al.*, 2011; Johnson, 2015).

Delineating the timing and geographic extent of increased environmental suitability for elevated NCV concentration is important for determining measures aimed at reducing the risk of exposure to NCV and the associated disease burden. To address the challenges of obtaining sufficient *in situ* environmental data in a timely manner, a significant amount of work has been devoted to developing EO technologies that can provide the data required for the development of early warning systems (Paz *et al.*, 2007; Lobitz *et al.*, 2000; Martinez-Urtaza *et al.*, 2008; Semenza *et al.*, 2017).

SST and SSS as remote sensing proxies have been tested and appear most promising for estimating or monitoring favorable environmental conditions for NCV occurrence and density (Phillips *et al.*, 2007; Zimmerman *et al.*, 2007; Baker-Austin *et al.*, 2010, 2013). These proxies have been used to estimate intra-seasonal variations in Vp density or abundance (Zimmerman *et al.*, 2007) as indicators of risk of contamination of water and shellfish (Phillips *et al.*, 2007; Konrad *et al.*, 2017), and as predictors of infectious disease risk (Baker-Austin *et al.*, 2013).

Challenges and questions

Challenges associated with obtaining remote sensing data include:

1. Complex system involving environmental factors at multiple scales

Working at multiple spatial and temporal scales – like a small coastal bay and a significantly larger coastal area – in a constant manner over large areas and prolonged time periods can make assessing the association of NCV with environmental factors a challenge.

2. Potentially significant technical requirements

The environmental factors identified in the previous section might necessitate the use of different sensors to obtain data with sufficient spatial, temporal, and spectral resolution. For example, a space-based open ocean salinity sensor cannot map closer than a few hundred kilometers from the coast and there are no immediate satellite sensor deployments for measuring coastal SSS. Similarly, satellite-based wind products are not reliable close to the coast, although high-to-moderate-resolution satellite SAR images are being used on a regular basis. Few optical and radar satellite sensors can provide the high spatial and temporal resolution data essential to study dynamic coastal and tidal regimes that change at hourly increments (Grimes *et al.*, 2014). Heavy precipitation events can decrease the salinity in estuaries and coastal areas and result in NCV growth. For example, *V. parahaemolyticus* increased at beaches subsequent to storm events by 155.7% at two distinct beaches in northern China (He *et al.*, 2019). However, such meteorological events are difficult to measure with remote sensing technology.

3. Knowledge gaps

Gaps in the understanding of biological and physical relationships between NCV and the above-mentioned environmental factors may undermine prioritization of data acquisition and interpretation. Moreover, NCV are not reportable in many parts of the world, which makes validation of these environmental factors difficult.

Based on these challenges, the following questions on critical issues and geospatial data requirements were submitted to a panel of experts during the One Earth – One Health Workshop on Contributions of Earth Observation to Public Health Practices:

- What EO and geospatial data are already available to detect and/or quantify the environmental factors that have a probable/possible influence on NCV in aquatic systems?
- What EO technologies are already in place that could provide the necessary information?
- What future optical and radar sensors might be able to provide the tools to collect EO and geospatial data for the purposes of better understanding the impacts of climate change on NCV in aquatic systems?

Responses and options

Expert consultation on these questions generated the following responses and options:

- Critical environmental factors measured by satellite sensors are SST and SSS; *in situ* data (e.g. from moorings, ships, buoys) can complement satellite data collection on these factors.
- EO and geospatial data already available for coastal and open ocean areas require relatively high and medium–low spatial resolution, respectively. They are collected by the Department of Fisheries and Oceans (DFO) in Canada and are also available publicly (e.g. from NOAA/CoastWatch, NASA,

Operational Sea Surface Temperature and Sea Ice Analysis [OSTIA]).

- Buoys offer data with high temporal resolution, whereas the temporal resolution of data from satellite sensors is moderate; in both cases the data are free.
- Significant EO systems and technologies, including radar, optical sensors, and infrared radiometers, are already in place for coastal and open ocean remote sensing of wind, waves and currents, ocean color, and SST, and for model development and adaptation (NOAA, E3).
- New and upcoming sources include the European Sentinel systems (radar and multispectral imaging), NOAA/NASA Joint Polar Satellite System (JPSS) Series Satellites, and the RADARSAT Constellation Mission, with high spatial and moderate temporal resolution and the prospect of less complex EO data analysis requirements.

The example of the ECDC Vibrio Map Viewer developed by ECDC is presented in the following sections to illustrate the benefits of using remote sensing data to predict human health risks associated with NCV and to highlight technological requirements (Semenza *et al.*, 2017).

Application example: The ECDC Vibrio Map Viewer from the European Environment and Epidemiology (E3) Geoportal

In and around the Baltic Sea and the eastern North Sea area from 1977–2010, a total of 283 NCV cases were reported, with the majority of these cases (234, or 85%) reported from 1997 onwards (Baker-Austin *et al.*, 2013). Over this period, the number of cases increased in correlation with temperature increases, with periods of reported infections closely associated with areas of maximum warming. Highly significant statistical association was found between the annual number of human cases and mean summer SST increase (Semenza *et al.*, 2017).

A pilot suitability model based on Baker-Austin *et al.*'s article (2013) on ocean warming and the risk of *Vibrio* was developed using salinity and SST to estimate the environmental suitability for NCV in coastal waters (Semenza *et al.*, 2017). The output of the model delineates coastal areas with environmental conditions suitable for the occurrence of human pathogenic NCV species that can drive the emergence of infections. The ECDC Vibrio Map Viewer is an EO early warning system for public health created to help reduce human exposure to contaminated coastal waters and consequently reduce the disease burden of NCV (Semenza *et al.*, 2017). By using EO and SST records, the influence of recent warming trends on the emergence and dynamics of vibriosis in the Baltic was assessed (Ebi *et al.*, 2017; Semenza *et al.*, 2017). The assessment analyzed epidemiological data together with long-term SST records and shorter-term data from NOAA that integrates satellite SST retrievals. A satellite-independent data set was used to confirm the results from the long-term SST records, especially those related to the strong summer warming trends detected in the Baltic Sea during the past three decades when remote sensing data became available (Baker-Austin *et al.*, 2013).

The ECDC Vibrio Map Viewer from the European Environment and Epidemiology (E3) network has been providing NCV environmental suitability maps on their website through the E3 Geoportal since 2013.[1] The ECDC Vibrio Map Viewer provides global environmental suitability maps for NCV that are based on a real-time model that has been calibrated to the Baltic Region in Northern Europe (Semenza *et al.*, 2017). It uses daily updated remote sensing data to provide information about environmentally suitable conditions such as salinity and SST for *Vibrio* spp. The model can also be used and calibrated for any other region of the world (Fig. 2.4.1).

Outcomes and benefits

What does the ECDC Vibrio Map Viewer do? The ECDC Vibrio Map Viewer identifies environmental suitability for optimal growth conditions for NCV with a near-real-time model that uses daily updated information from EO data on water temperature and salinity (Semenza *et al.*, 2017). In light of increasing SST in the Baltic Sea due to climate change, the environmental suitability

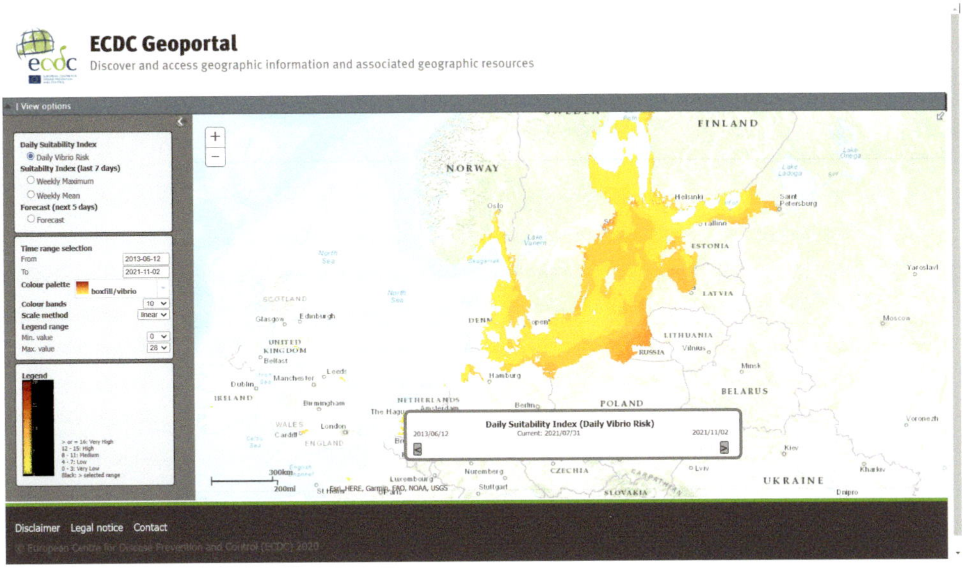

Fig. 2.4.1. Sample display of the daily NCV risk map viewer provided by the ECDC Vibrio Map Viewer from the E3 (ECDC18) network for the Baltic Sea on 31 July 2021. Areas depicted in dark orange indicate high to very high risk and areas in yellow indicate low to very low risk.

has been projected forward in time to delineate coastal areas at risk for NCV growth in the future (Lindgren *et al.*, 2012; Ebi *et al.*, 2017; Semenza *et al.*, 2017).

The model generates a daily map and 5-day forecasts of the environmental suitability of NCV. The map viewer categorizes the environmental suitability for NCV growth, as low, very low, medium, high, and very high.

How are the findings used? The epidemic intelligence team at ECDC monitors the Vibrio Map Viewer on a daily basis during the summer months in order to identify coastal areas with increased environmental suitability. The findings are reported bi-weekly in the ECDC Communicable Disease Threats Report (CDTR), which is distributed to national state epidemiologists in Europe (ECDC, 2019). The CDTR also discusses options for public health prevention and control actions. For example, these might include beach closures, issuing alerts when the environmental suitability for NCV infections is predicted to increase, notifying health care providers, and encouraging at-risk individuals (e.g. children, the elderly, cancer patients, and immune-compromised individuals) to avoid recreational water use to limit exposure to NCV.

Delineating areas of environmental suitability for NCV in the near future can inform public health decision making. For example, in 2018, the model detected a significant increase in the geographic extent of environmental suitability for NCV in coastal areas of the Baltic Sea and the Gulf of Bothnia, using both daily and forecasted values. This was linked to abnormally high temperatures and drought conditions in several countries around the Baltic Sea. At the beginning of August 2018, ECDC sent a notice to state epidemiologists around the Baltic Sea through the Epidemic Intelligence Information System for Food-and Waterborne Diseases and Zoonoses (EPIS FWD), alerting them to the possibility of an increase in NCV cases.

Technical requirements

What EO data are needed?

- SSS and SST

The ECDC Vibrio Map Viewer uses daily updated remotely sensed SST and SSS data, *in situ* data, and climatological data for coastal waters. For the Baltic Sea, SSS demarcates the regions suitable for NC infections (Copernicus Marine Environment Monitoring Service, 2019) and SST serves as a risk predictor (Semenza *et al.*, 2017). The estimates for SST and SSS were obtained from different satellites and models from the following sources:

- (SST) US Department of Commerce (USDOC); NOAA; National Environmental Satellite Data and Information Service (NESDIS); and Geo-Polar Blended Global SST Level (Imagers + AVHRR + VIIRS). These fields integrate data from US, Japanese, and European geostationary infrared imagers and from polar infrared (US and European) SST sensors to produce a high-resolution 5-km product.
- (SSS) NOAA/NCEP (National Centers for Environmental Prediction) Global Real-Time Ocean Forecast System.
- (SSS) Navy Coastal Ocean Model (NCOM) for the Gulf of Mexico, Caribbean, and US East Coast.
- (SSS) Operational Mercator Global Ocean Analysis and Forecast System.
- (SSS) Iberian Biscay Irish (IBI) Ocean Analysis and Forecasting System.
- (SSS) Forecasting Ocean Assimilation Model (FOAM); 7 km Atlantic Margin model (AMM7).
- (SSS) Baltic Sea Physical Analysis and Forecasting Product.
- (SSS) Mediterranean Sea Physics Analysis and Forecast.
- (SSS) Black Sea Physics Analysis and Forecast.

The model's SST estimates are used for the short-term forecast and are replaced by the satellite-based SST as they are available. They also serve as a backup in case of (rare) delays in the primary SST data set. The daily map integrates all these data sets to obtain the environmental suitability index using threshold values of 18°C for SST and 28 Practical Salinity Units (psu) for SSS. The nominal spatial resolution of the output is 5 km. Future improvements will add more data sets to the process, especially in areas where global models do not produce accurate representation of ocean dynamics (e.g. Gulf of Guinea), and will include epidemiological data to refine the model and improve the overall quality of the results.

Cumulative fields

Daily estimates provide inputs to compute weekly environmental suitability. This is used to generate the:

- cumulative weekly index (mean value) using data from the past 7 days ("CWIMean"); and
- cumulative weekly index (maximum value) using data from the past 7 days ("CWIMax").

The cumulative weekly index is generated on a daily basis, using the same inputs referenced above and the same procedures as those used to estimate daily risk.

Short-term forecast

The model can also generate short-term forecasts, which are created using the same algorithms and data structures, and which can be used to inform planning and health impact assessments. ECDC also conducts continuous evaluation of the quality of forecasts by comparing them with observations and analyzed results.

What resources are needed? During the past decade, GIS technology has been widely used for purposes ranging from natural resource protection to archeological research, from fish stock assessment to weather monitoring. The ECDC Vibrio Map Viewer uses a suite of spatio-temporal products that, through interoperable techniques, can access and use remote layers via OGC Web Services (OWS).[3] Using this approach, it is possible to combine local and remote raster and vector layers, change the projection, display and customize plots, retrieve data, and incorporate the desired layers within a specific GIS environment. Based on XML schema, our model facilitates the creation of customized interfaces that adjust content to user needs. These interfaces can be easily deployed at minimum cost and can be used for coastal and meso- and large-scale applications.

The ECDC Vibrio Map Viewer, which closely integrates with NOAA/CoastWatch's Ocean-Viewer system, generates a Vibrio Risk Index and the mechanisms to distribute fields with improved data access and visualization and with the ability to share multiple data sets, to combine digital data from vector and raster sources, and to serve both operational and non-operational objectives. Further advantages are that it is a modular, standard-driven approach that allows for further technological development, it uses existing infrastructure, it provides access to timely and accurate information, it implements online and individualized access to services and products, it serves as a model of synergy between leading edge information technologies, and it benefits research and operational programs.

The outputs of the various procedures comprising the risk model are originally encoded in netCDF format. This open standard, intended to improve sharing of scientific data and metadata, is currently a *de facto* standard that is widely used to store remote sensing and model data. It provides a complete API for software development and multiple features (e.g. chunking, deflation, machine independent, simple data model, user data types, unlimited dimensions), which can greatly improve read/write operations, overall performance, and interoperability. The files are made compatible with Climate and Forecast (CF) metadata convention, which provides an extensible controlled vocabulary for many of the scientific parameters being used by the community. Our netCDF files are therefore self-describing, packing all data and metadata together.

The distribution scheme uses various protocols and services. Among them, OWS allow efficient processing of geospatial data and promote automatic exploitation. They provide simple interfaces for requesting georeferenced products that could be stored in different geospatial databases.

The ECDC Vibrio Map Viewer uses two services: Web Map Service (WMS) and Web Coverage Service (WCS).[4] The ESRI Silverlight toolkit can display WMS GetMap requests and, consequently, risk maps can be easily integrated into the E3 Geoportal I.2 viewer. ArcGIS, through ArcCatalog and ArcMap, can also act as a client for WCS servers, allowing geographical data requests using platform-independent interfaces.

WMS: This is a standard to produce georeferenced maps. Two WMS servers are available, one of which has been implemented with additional functions not included in the specification (e.g. GetMetadata Method). It supports the following methods:

- GetCapabilities (to obtain service metadata);
- GetMap (to obtain a map);
- GetFeatureInfo (information about queryable features);
- GetLegendGraphics (to acquire a legend); and
- GetMetadata (to provide metadata not included in the standard GetCapabilities document).

WCS: This service serves gridded data in various formats (GeoTIFF and netCDF4). There are three types of requests: GetCapabilities (delivers service properties), DescribeCoverage (specific information about coverage offerings), and GetCoverage (to retrieve the data). An example of the GetCoverage request can be downloaded here.[5]

Future perspectives

What further developments are needed? Although the factors that most influence the occurrence and abundance of NCV in coastal waters are SST and SSS, NCV ecology and growth also depend on other variables like marine nutrient concentrations, river discharge, and algae blooms (Julie *et al.*, 2010; Johnson *et al.*, 2012; Boer *et al.*, 2013). Thus, the environmental suitability shown by the ECDC Vibrio Map Viewer represents only an approximation of actual suitability and local variation. Adding these other variables to the model might improve sensitivity, specificity, and positive predictive value, although this would need to be confirmed by epidemiological data.

The models used in the near future for inferring/forecasting risk (rather than ecological suitability) will be based on artificial intelligence using an artificial neural network. These approaches allow for the integration of data from disparate sources – environmental variables, epidemiological data, pathogen population dynamics, genomics, disease outcomes, and exposure – in order to obtain more accurate proxies of environmental scenarios of risk. As this methodology incorporates a "training" process using real data, outcomes from the model can be tested based on real scenarios. However, access to such data might be difficult. For example, data on exposure during recreational water use are often hard to come by but are available for certain coastal regions (Dwight *et al.*, 2007).

Sensors such as VIIRS (on board JPSS satellites) and Ocean and Land Colour Instrument (OLCI) (Sentinel-3) could provide high-quality estimates of these parameters; for example, information on turbidity naturally present in coastal water or from other sources such as river discharge (e.g. suspended particulate matter [SPM], colored dissolved organic matter [CDOM]) and on algae or phytoplankton blooms (e.g. chlorophyll-a [Chla]). However, *in situ* ecological data such as from buoys are not always available and the cost and time associated with collecting ecological *in situ* data need to be considered. In view of these potential limitations, it would be valuable to first assess the predictive value of data on these variables in the exposure response relationship.

Obtaining SSS data can be a challenge in coastal areas that do not have a continuous operation instrument for this type of data. Current remote sensing data that are able to estimate surface salinity are not suitable for coastal systems since their spatial resolution is too coarse (>25 km), which generates spectral mixing between water and land. A sensor or methodology capable of estimating the surface salinity of coastal water on a daily basis would provide valuable data for future early warning systems for NCV.

Better data are needed in Europe on the abundance and persistence of NCV in the environment and on the number of cases of infection. Although NCV infection can lead to serious health consequences, only a few countries in the region, such as Sweden, have national epidemiological data. Analysis of Swedish *Vibrio* cases includes information on county, statistical date, and onset of disease, type of infection, *Vibrio* spp., serotype, transmission pathway, sex, and age group of the case. This could then be correlated with data on environmental suitability for NCV and help to calibrate the scale of the ECDC Vibrio Map Viewer for a specific geographic region.

Furthermore, it is not always possible to link human cases of infection to national or international exposure, which introduces a certain level of exposure misclassification. More accurate documentation of infection source, together with the collection of data on human cases, would greatly enhance the value of the Vibrio Map Viewer and efforts to prevent and control NCV infections.

What other opportunities do EO technologies and data offer?

Remote sensing of ocean biota is performed globally through passive (multispectral, hyperspectral) or active (SAR, Radar) methods. Passive optical sensors capture the backscattered light from below the water surface. Infrared passive sensors detect thermal variations at the sea surface (SST). Multispectral sensors could also provide the opportunity to estimate potential NCV proxies such as SPM, CDOM, and Chla as well as providing their inherent optical properties (Table 2.4.1).

Hyperspectral remote sensing offers an effective approach for frequent, synoptic water quality measurements over a large spatial area. However, the optical complexity of coastal waters makes water quality monitoring by remote sensing in estuarine and littoral waters a challenge. The use of hyperspectral sensing for the diagnosis of ecological issues is a tool that is widely used (Fan, 2014), and one of the proven benefits of this approach is the ability to detect early specific spectral changes. This attribute of hyperspectral sensing helps to identify early optical changes in water constituents such as harmful algal blooms and the possibility of bacterial outbreaks. Satellites with hyperspectral detectors have 1 nm (or less) wide bands that usually span the spectrum from 400 to 700 nm to measure light emanating from the sea surface at these specific wavelengths. Algorithms could be developed to assess the relationships between the amount of light at specific wavelengths and estimated concentrations of the various constituents such as chlorophyll, particles, CDOM, or potential indicators of NCV presence and/or abundance.

Active sensors carry their own lighting source that is beamed down and the reflected beam carries information about the biogeochemical state of water. Active sensors work mainly within microwave spectra and can collect valuable information on sea surface roughness and winds and can detect and track oil spills and ice (Passive Active L- and S-band Sensor, PALS project, www.nasa.gov). They could also be used for gathering information on bacteria such as NCV.

Table 2.4.1. Possibilities for EO technologies, satellite sensors, and potential proxies or indicators of *Vibrio* occurrence in coastal waters.

Remote sensing technology	Examples of satellite sensors	Examples of proxies/ indicators	References
Multispectral and future hyperspectral	MODIS VIIRS Landsat-8 MSI (Sentinel-2) OLCI (Sentinel-3) SeaWiFS Future hyperspectral missions: PACE (NASA) WaterSat (CSA)	Suspended particulate matter (SMP) Chlorophyll-a (Chla) Colored dissolved organic matter (CDOM) Coefficient of attenuation (Kd) Inherent optical properties	Gilerson *et al.*, 2010; Johnson *et al.*, 2012; Escobar *et al.*, 2015
Thermal infrared	AVHRR MODIS VIIRS Geostationary sensors: (ABI/ AHI/SEVERI) SLSTR (Sentinel-3)	Sea surface temperature	Lobitz *et al.*, 2000; Martinez-Urtaza *et al.*, 2012; Akbari *et al.*, 2017
Synthetic aperture radar (SAR) or scatterometer	RADARSAT-2 Sentinel-1 TerraSAR-X QuickScat Future SAR mission RCM (CSA)	Roughness of the sea Waves (wind speed)	Lobitz *et al.*, 2000
Radar altimetry	Jason-1/-2/-3 Cryosat-2 SARAL/AltiKa Sentinel-3	Topography of sea Geostrophic currents	Martin-Puig *et al.*, 2016

ABI, Advanced Baseline Imager; AHI, Advanced Himawari Imager; AltiKa, high-resolution altimeter including bi-frequency radiometric function; AVHRR, advanced very high-resolution radiometer; CSA, Canadian Space Agency; SLSTR, Sea and Land Surface Temperature Radiometer; MODIS, Moderate Resolution Imaging Spectroradiometer; MSI, multispectral instrument; OLCI, Ocean and Land Colour Instrument; PACE, Plankton, Aerosol, Cloud, ocean Ecosystem; RCM, RADARSAT Constellation Mission; SARAL, Satellite with Argos and AltiKa; SeaWiFS, Sea-viewing Wide Field-of-view Sensor; SEVIRI, Spinning Enhanced Visible and Infrared Imager; VIIRS, Visible/Infrared Imager Radiometer Suite.

Perspectives for the ECDC Vibrio Map Viewer

Future studies should validate the ECDC Vibrio Map Viewer in different geographic settings with relevant epidemiologic data. Validation studies like these could also study the connection between SST/SSS EO data and the abundance and persistence of NVC in the environment. A technical challenge for EO are extreme precipitation events that can temporarily reduce SSS in estuaries and increase the environmental suitability of NVC growth (He *et al.*, 2019). Detecting and predicting such events with EO would serve public health well. It would also be desirable to apply the concept of the ECDC Vibrio Map Viewer to other health endpoints. However, this would require a better understanding of the underlying drivers of disease transmission. In many cases, there are multiple drivers of transmission that are not dependent on a single environmental/climatic variable. For example, *Campylobacter* is not capable of reproducing outside of an animal host (e.g. poultry, pigs) and the seasonal incidence peak does not always occur during the hottest time of the year, which indicates the importance of other drivers of transmission.

Conclusion

NCV infections in humans can cause mild, self-limiting gastroenteritis but also wound infections that can rapidly result in septicemia and necrotizing fasciitis with a high fatality rate. Climate change will disproportionally affect coastal regions in northern and southern latitudes and is likely to greatly expand the geographic range of NCV. Indeed, the number of wound infections from exposure to NCV during recreational water use observed in temperate zones has steadily increased during the past few decades due to climate change. Similarly, shellfish-associated food-borne outbreaks due to NCV at high latitudes have also increased due to warming. Since NCV thrive in warm and brackish marine environments these trends are projected to continue. However, there is little epidemiologic surveillance of infections due to NCV since they are not reportable in many countries. Thus, EO services can be used as proxies for the lack of epidemiologic data and can monitor the environmental precursors of this disease.

We show that forecasts of SST from EO can be used as climatic indicators of future *Vibrio* growth in coastal regions of the Baltic Sea, as part of an early warning system. Early detection and response to climate-sensitive pathogens can reduce the burden of disease better than passive surveillance systems (Morin *et al.*, 2018). However, this system relies on EO monitoring of SST and other environmental conditions (salinity) at scales useful for implementation of interventions and timely sharing of the information with public health professionals. Areas deemed suitable for NCV growth can then be identified and beaches can be closed and other measures taken to minimize human exposure to NCV. Ultimately, these EO systems for coastal monitoring are intended to reduce the burden of disease in human populations. However, system performance needs continuous evaluation due to its interconnected nature in order to maximize efficiency and improve efficacy.

The lack of empirical NVC data, both epidemiologic and environmental, makes EO an attractive alternative to traditional methods that are not applicable at the same scale in time and space. Monthly projections predict that SST suitability for *Vibrio* in the Baltic Sea will increase due to climate change. In particular, a marked upward trend is projected for SST during July, August, and September, but even more so during the months immediately before and after summer (June and October). This early warning system based on EO can be regarded as a climate change adaptation in a warming climate. In fact, SST and vibriosis are currently being used as indicators for the Lancet Countdown on health and climate change (Watts *et al.*, 2018).

Notes

[1] https://e3geoportal.ecdc.europa.eu/SitePages/Vibrio%20Map%20Viewer.aspx (accessed 4 January 2022).
[2] https://e3geoportal.ecdc.europa.eu/SitePages/Home.aspx (accessed 4 January 2022).
[3] https://www.opengeospatial.org/standards/owc (accessed 4 January 2022).
[4] For additional information, see http://www.opengeospatial.org/standards (accessed 4 January 2022).
[5] https://cwcgom.aoml.noaa.gov/thredds/wcs/VIBRIO_RISK/RISK.nc?service=WCS&version=1.0.0&request=GetCoverage&format=GeoTIFF&coverage=daily_vibrio_risk&TIME=2013-09-01T12:00:00.000Z&elevation=0&bbox=-20,20,60,65 (accessed 4 January 2022).

References

Akbari, E., Alavipanah, K.S., Jeihouni, M., Hajeb, M., Haase *et al.* (2017) A review of ocean/sea subsurface water temperature studies from remote sensing and non-remote sensing methods. *Water* 9, 1–25. DOI: https://doi.org/10.3390/w9120936.

Ali, M., Nelson, A.R., Lopez, A.L., and Sack, D.A. (2015) Updated global burden of cholera in endemic countries. *PLoS Neglected Tropical Diseases* 9. DOI: https://doi.org/10.1371/journal.pntd.0003832.

Amel, B.K., Amine, B., and Amina, B. (2008) Survival of *Vibrio fluvialis* in seawater under starvation conditions. *Microbiological Research* 163, 323–328. DOI: https://doi.org/10.1016/j.micres.2006.06.006.

Baker-Austin, C., Stockley, L., Rangdale, R., and Martinez-Urtaza, J. (2010) Environmental occurrence and clinical impact of *Vibrio vulnificus* and *Vibrio parahaemolyticus*: A European perspective. *Environmental Microbiology Reports* 2, 7–18. DOI: https://doi.org/10.1111/j.1758-2229.2009.00096.x.

Baker-Austin, C., Trinanes, J.A., Taylor, N.G.H., Hartnell, R., Siitonen, A. *et al.* (2013) Emerging *Vibrio* risk at high latitudes in response to ocean warming. *Nature Climate Change* 3, 73–77. DOI: https://doi.org/10.1038/nclimate1628.

Baker-Austin, C., Trinanes, J.A., Gonzalez-Escalona, N., and Martinez-Urtaza, J. (2017) Non-cholera *Vibrios*: The microbial barometer of climate change. *Trends in Microbiology* 25, 76–84. DOI: https://doi.org/10.1016/j.tim.2016.09.008.

Boer, S.I., Heinemeyer, E.A., Luden, K., Erler, R., Gerdts, G. *et al.* (2013) Temporal and spatial distribution patterns of potentially pathogenic *Vibrio* spp. at recreational beaches of the German North Sea. *Microbial Ecology* 65, 1052–1067. DOI: https://doi.org/10.1007/s00248-013-0221-4.

Ceccarelli, D., Hasan, N.A., Huq, A., and Colwell, R.R. (2013) Distribution and dynamics of epidemic and pandemic *Vibrio parahaemolyticus* virulence factors. *Frontiers in Cellular and Infection Microbiology* 3, 1–9. DOI: https://doi.org/10.3389/fcimb.2013.00097.

Centers for Disease Control and Prevention (CDC) (2014) National Enteric Disease Surveillance: Cholera and Other *Vibrio* Illness Surveillance (COVIS) Annual Summary. Available at: https://www.cdc.gov/vibrio/surveillance.html (accessed 14 June 2021).

Copernicus Marine Environment Monitoring Service (2019) Homepage. Available at: http://marine.copernicus.eu/ (accessed 14 June 2021).

Coutard, F., Crassous, P., Droguet, M., Gobin, E., Colwell, R. *et al.* (2007) Recovery in culture of viable but nonculturable *Vibrio parahaemolyticus*: Regrowth or resuscitation? *The ISME Journal* 1, 111–120. DOI: https://doi.org/10.1038/ismej.2007.1.

DePaola, A., Nordstrom, J.L., Bowers, J.C., Wells, J.G., and Cook, D.W. (2003) Seasonal abundance of total and pathogenic *Vibrio parahaemolyticus* in Alabama oysters. *Applied and Environmental Microbiology* 69, 1521–1526. DOI: https://doi.org/10.1128/AEM.69.3.1521-1526.2003.

Dwight, R.H., Brinks, M.V., SharavanaKumar, G., and Semenza, J.C. (2007) Beach attendance and bathing rates for southern California beaches. *Ocean & Coastal Management* 50, 847–858. DOI: https://doi.org/10.1016/j.ocecoaman.2007.04.002.

Ebi, K.L., Ogden, N.H., Semenza, J.C., and Woodward, A. (2017) Detecting and attributing health burdens to climate change. *Environmental Health Perspectives* 125, 085004. DOI: https://doi.org/10.1289/EHP1509.

Escobar, L.E., Ryan, S.J., Stewart-Ibarra, A.M., Finkelstein, J.L., King, C.A. *et al.* (2015) A global map of suitability for coastal *Vibrio cholerae* under current and future climate conditions. *Acta Tropica* 149, 202–211. DOI: https://doi.org/10.1016/j.actatropica.2015.05.028.

European Centre for Disease Prevention and Control (ECDC) (2019) Communicable disease threats reports. Available at: https://ecdc.europa.eu/en/threats-and-outbreaks/reports-and-data/weekly-threats (accessed 14 June 2021).

Fan, C. (2014) Spectral analysis of water reflectance for hyperspectral remote sensing of water quality in estuarine water. *Journal of Geoscience and Environment Protection* 02, 19–27. DOI: https://doi.org/10.4236/gep.2014.22004.

Gilerson, A.A., Gitelson, A.A., Zhou, J., Gurlin, D., Moses, W. *et al.* (2010) Algorithms for remote estimation of chlorophyll-a in coastal and inland waters using red and near infrared bands. *Optics Express* 18, 24109–24125. DOI: https://doi.org/10.1364/OE.18.024109.

Gonzalez, D.J., Gonzalez, R.A., Froelich, B.A., Oliver, J.D., Noble, R.T. *et al.* (2014) Non-native macroalga may increase concentrations of *Vibrio* bacteria on intertidal mudflats. *Marine Ecology Progress Series* 505, 29–36. DOI: https://doi.org/10.3354/meps10771.

Government of Canada (2015) National Enteric Surveillance Program Annual Summary 2013. *Public Health Agency of Canada*, Guelph. Available at: https://publications.gc.ca/collections/collection_2016/aspc-phac/HP37-15-2013-eng.pdf (accessed 21 June 2021).

Grimes, D.J., Ford, T.E., Colwell, R.R., Baker-Austin, C., Martinez-Urtaza, J. *et al.* (2014) Viewing marine bacteria, their activity and response to environmental drivers from orbit: Satellite remote sensing of bacteria. *Microbial Ecology* 67, 489–500. DOI: https://doi.org/10.1007/s00248-013-0363-4.

He, Y., He, Y., Sen, B., Li, H., Li, J. *et al.* (2019) Storm runoff differentially influences the nutrient concentrations and microbial contamination at two distinct beaches in northern China. *Science of the Total Environment* 663, 400–407. DOI: https://doi.org/10.1016/j.scitotenv.2019.01.369.

Hlady, W.G. and Klontz, K.C. (1996) The epidemiology of *Vibrio* infections in Florida, 1981–1993. *Journal of Infectious Diseases* 173, 1176–1183. DOI: https://doi.org/10.1093/infdis/173.5.1176.

Janda, J.M., Powers, C., Bryant, R.G., and Abbott, S.L. (1988) Current perspectives on the epidemiology and pathogenesis of clinically significant *Vibrio* spp. *Clinical Microbiology Reviews* 1, 245–267. DOI: https://doi.org/10.1128/cmr.1.3.245.

Johnson, C.N. (2015) Influence of environmental factors on *Vibrio* spp. in coastal ecosystems. *Microbiology Spectrum* 3, 1–18. DOI: https://doi.org/10.1128/microbiolspec.VE-0008-2014.

Johnson, C.N., Flowers, A.R., Noriea, N.F., Zimmerman, A.M., Bowers, J.C. *et al.* (2010) Relationships between environmental factors and pathogenic *Vibrios* in the northern Gulf of Mexico. *Applied and Environmental Microbiology* 76, 7076–7084. DOI: https://doi.org/10.1128/AEM.00697-10.

Johnson, C.N., Bowers, J.C., Griffitt, K.J., Molina, V., Clostio, R.W. *et al.* (2012) Ecology of *Vibrio parahaemolyticus* and *Vibrio vulnificus* in the coastal and estuarine waters of Louisiana, Maryland, Mississippi, and Washington (United States). *Applied and Environmental Microbiology* 78, 7249–7257. DOI: https://doi.org/10.1128/AEM.01296-12.

Julie, D., Solen, L., Antoine, V., Jaufrey, C., Annick, D. *et al.* (2010) Ecology of pathogenic and non-pathogenic *Vibrio parahaemolyticus* on the French Atlantic coast. Effects of temperature, salinity, turbidity and chlorophyll a. *Environmental Microbiology* 12, 929–937. DOI: https://doi.org/10.1111/j.1462-2920.2009.02136.x.

Kaneko, T. and Colwell, R.R. (1973) Ecology of *Vibrio parahaemolyticus* in Chesapeake Bay. *Journal of Bacteriology* 113, 24–32. DOI: https://doi.org/10.1128/JB.113.1.24-32.1973.

Kelly, M.T. and Stroh, E.M. (1988) Temporal relationship of *Vibrio parahaemolyticus* in patients and the environment. *Journal of Clinical Microbiology* 26, 1754–1756. DOI: https://doi.org/10.1128/jcm.26.9.1754-1756.1988.

Kokashvili, T., Whitehouse, C.A., Tskhvediani, A., Grim, C.J., Elbakidze, T. *et al.* (2015) Occurrence and diversity of clinically important *Vibrio* species in the aquatic environment of Georgia. *Frontiers in Public Health* 3, 1–12. DOI: https://doi.org/10.3389/fpubh.2015.00232.

Konrad, S., Paduraru, P., Romero-Barrios, P., Henderson, S.B., and Galanis, E. (2017) Remote sensing measurements of sea surface temperature as an indicator of *Vibrio parahaemolyticus* in oyster meat and human illnesses. *Environmental Health* 16, 1–11. DOI: https://doi.org/10.1186/s12940-017-0301-x.

Lindgren, E., Andersson, Y., Suk, J.E., Sudre, B., and Semenza, J.C. (2012) Monitoring EU emerging infectious disease risk due to climate change. *Science* 336, 418–419. DOI: https://doi.org/10.1126/science.1215735.

Lobitz, B., Beck, L., Huq, A., Wood, B., Fuchs, G. *et al.* (2000) Climate and infectious disease: Use of remote sensing for detection of *Vibrio cholerae* by indirect measurement. *Proceedings of the National Academy of Sciences of the United States of America* 97(4), 1438–1443. DOI: https://doi.org/10.1073/pnas.97.4.1438.

Martin-Puig, C., Leuliette, E., Lillibridge, J., and Roca, M. (2016) Evaluating the performance of Jason-2 open-loop and closed-loop tracker modes. *Journal of Atmospheric and Oceanic Technology* 33, 2277–2288. DOI: https://doi.org/10.1175/JTECH-D-16-0011.1.

Martinez-Urtaza, J., Huapaya, B., Gavilan, R.G., Blanco-Abad, V., Ansede-Bermejo, J. *et al.* (2008) Emergence of Asiatic Vibrio diseases in South America in phase with El Nino. *Epidemiology* 19, 829–837. DOI: https://doi.org/10.1097/EDE.0b013e3181883d43.

Martinez-Urtaza, J., Blanco-Abad, V., Rodriguez-Castro, A., Ansede-Bermejo, J., Miranda, A. *et al.* (2012) Ecological determinants of the occurrence and dynamics of *Vibrio parahaemolyticus* in offshore areas. *The ISME Journal* 6, 994–1006. DOI: https://doi.org/10.1038/ismej.2011.156.

McCarter, L. (1999) The multiple identities of *Vibrio parahaemolyticus*. *Journal of Molecular Microbiology and Biotechnology* 1, 51–57. DOI: https://doi.org/10.3390/ijerph15051048.

Morin, C.W., Semenza, J.C., Trtanj, J.M., Glass, G.E., Boyer, C. *et al.* (2018) Unexplored opportunities: Use of climate- and weather-driven early warning systems to reduce the burden of infectious diseases. *Current Environmental Health Reports* 5, 430–438. DOI: https://doi.org/10.1007/s40572-018-0221-0.

Motes, M.L., DePaola, A., Cook, D.W., Veazey, J.E., Hunsucker, J.C. *et al.* (1998) Influence of water temperature and salinity on *Vibrio vulnificus* in Northern Gulf and Atlantic Coast oysters (*Crassostrea virginica*). *Applied and Environmental Microbiology* 64, 1459–1465. DOI: https://doi.org/10.1128/AEM.64.4.1459-1465.1998.

Nigro, O.D., Hou, A., Vithanage, G., Fujioka, R.S., and Steward, G.F. (2011) Temporal and spatial variability in culturable pathogenic *Vibrio* spp. in Lake Pontchartrain, Louisiana, following Hurricanes Katrina and Rita. *Applied and Environmental Microbiology* 77, 5384–5393. DOI: https://doi.org/10.1128/AEM.02509-10.

Oliver, J.D., Hite, F., McDougald, D., Andon, N.L., and Simpson, L.M. (1995) Entry into, and resuscitation from, the viable but nonculturable state by *Vibrio vulnificus* in an estuarine environment. *Applied and Environmental Microbiology* 61, 2624–2630. DOI: https://doi.org/10.1128/aem.61.7.2624-2630.1995.

Paz, S., Bisharat, N., Paz, E., Kidar, O., and Cohen, D. (2007) Climate change and the emergence of *Vibrio vulnificus* disease in Israel. *Environmental Research* 103, 390–396. DOI: https://doi.org/10.1016/j.envres.2006.07.002.

Pfeffer, C.S., Hite, M.F., and Oliver, J.D. (2003) Ecology of *Vibrio vulnificus* in estuarine waters of eastern North Carolina. *Applied and Environmental Microbiology* 69, 3526–3531. DOI: https://doi.org/10.1128/AEM.69.6.3526-3531.2003.

Phillips, A.M., Depaola, A., Bowers, J., Ladner, S., and Grimes, D.J. (2007) An evaluation of the use of remotely sensed parameters for prediction of incidence and risk associated with *Vibrio parahaemolyticus* in Gulf Coast oysters (*Crassostrea virginica*). *Journal of Food Protection* 70, 87–884. DOI: https://doi.org/10.4315/0362-028x-70.4.879.

Sarkar, B.L., Nair, G.B., Banerjee, A.K., and Pal, S.C. (1985) Seasonal distribution of *Vibrio parahaemolyticus* in freshwater environs and in association with freshwater fishes in Calcutta. *Applied and Environmental Microbiology* 49, 132–136. DOI: https://doi.org/10.1128/AEM.49.1.132-136.1985.

Schijven, J., Bouwknegt, M., de Roda Husman, A.M., Rutjes, S., Sudre, B. *et al.* (2013) A decision support tool to compare waterborne and foodborne infection and/or illness risks associated with climate change. *Risk Analysis* 33, 2154–2167. DOI: https://doi.org/10.1111/risa.12077.

Semenza, J.C., Herbst, S., Rechenburg, A., Suk, J.E., Höser, C. *et al.* (2012a) Climate change impact assessment of food- and waterborne diseases. *Critical Reviews in Environmental Science and Technology* 42, 857–890. DOI: https://doi.org/10.1080/10643389.2010.534706.

Semenza, J.C., Höser, C., Herbst, S., Rechenburg, A., Suk, J.E. *et al.* (2012b) Knowledge mapping for climate change and food- and waterborne diseases. *Critical Reviews in Environmental Science and Technology* 42, 378–411. DOI: https://doi.org/10.1080/10643389.2010.518520.

Semenza, J.C., Trinanes, J., Lohr, W., Sudre, B., Lofdahl, M. *et al.* (2017). Environmental suitability of *Vibrio* infections in a warming climate: An early warning system. *Environmental Health Perspectives* 125(10), 107004. DOI: https://doi.org/10.1289/EHP2198.

Siboni, N., Balaraju, V., Carney, R., Labbate, M., and Seymour, J.R. (2016) Spatiotemporal dynamics of *Vibrio* spp. within the Sydney Harbour Estuary. *Frontiers in Microbiology* 7, 1–15. DOI: https://doi.org/10.3389/fmicb.2016.00460.

Taylor, M., Cheng, J., Sharma, D., Bitzikos, O., Gustafson, R. *et al.* (2018) Outbreak of *Vibrio parahaemolyticus* associated with consumption of raw oysters in Canada, 2015. *Foodborne Pathogens and Disease* 15, 554–559. DOI: https://doi.org/10.1089/fpd.2017.2415.

Thompson, J.R., Randa, M.A., Marcelino, L.A., Tomita-Mitchell, A., Lim, E. *et al.* (2004) Diversity and dynamics of a North Atlantic coastal *Vibrio* community. *Applied and Environmental Microbiology* 70, 4103–4110. DOI: https://doi.org/10.1128/AEM.70.7.4103-4110.2004.

Urquhart, E.A., Jones, S.H., Yu, J.W., Schuster, B.M., Marcinkiewicz, A.L. *et al.* (2016) Environmental conditions associated with elevated *Vibrio parahaemolyticus* concentrations in Great Bay Estuary, New Hampshire. *PLoS ONE* 11, 1–15. DOI: https://doi.org/10.1371/journal.pone.0155018.

Venkateswaran, K., Kiiyukia, C., Takaki, M., Nakano, H., Matsuda, H. *et al.* (1989) Characterization of toxigenic *Vibrios* isolated from the freshwater environment of Hiroshima, Japan. *Applied and Environmental Microbiology* 55, 2613–2618. DOI: https://doi.org/10.1128/aem.55.10.2613-2618.1989

Watts, N., Amann, M., Arnell, N., Ayeb-Karlsson, S., Belesova, K. *et al.* (2018) The 2018 report of the Lancet Countdown on health and climate change: Shaping the health of nations for centuries to come. *The Lancet* 392, 2479–2514. DOI: https://doi.org/10.1016/S0140-6736(18)32594-7.

Williams, L.A. and Larock, P.A. (1985) Temporal occurrence of *Vibrio* species and *Aeromonas hydrophila* in estuarine sediments. *Applied and Environmental Microbiology* 50, 1490–1495. DOI: https://doi.org/10.1128/aem.50.6.1490-1495.1985.

Zhong, L., Chen, J., Zhang, X.-H., and Jiang, Y.-A. (2009) Entry of *Vibrio cincinnatiensis* into viable but nonculturable state and its resuscitation. *Letters in Applied Microbiology* 48, 247–252. DOI: https://doi.org/10.1111/j.1472-765X.2008.02522.x.

Zimmerman, A.M., DePaola, A., Bowers, J.C., Krantz, J.A., Nordstrom, J.L. *et al.* (2007) Variability of total and pathogenic *Vibrio parahaemolyticus* densities in northern Gulf of Mexico water and oysters. *Applied and Environmental Microbiology* 73, 7589–7596. DOI: https://doi.org/10.1128/AEM.01700-07.

2.5 Vulnerable Populations

Marion Borderon[1], Antoinette Ludwig[2], and Stéphanie Brazeau[2]
[1]*Department of Geography and Regional Research, University of Vienna, Vienna, Austria;* [2]*Public Health Risk Sciences Division, National Microbiology Laboratory, Public Health Agency of Canada, Saint-Hyacinthe, Canada*

Context

Human vulnerability to diseases is the key measure of the potential impacts of a given change on the well-being of human populations. Human *vulnerability* is most often defined by the complex interaction between the *susceptibility* of an individual, community, or population, their *exposure* to a threat (i.e. the hazard), and their capacity for *resilience* (Turner, 2010). *Exposure* can be expressed by the presence or proximity of and the potential contact between the hazard and susceptible populations (Blaikie *et al.*, 1994; Cutter, 1996). *Susceptibility* can include immune capacity, behavior, or perception of a person or a community facing a threat (Fritzsche *et al.*, 2014). *Coping capacity* constitutes resilience, expressed in a public health context as the human ability to face a threat and develop coping skills to protect or to recover from diseases. These capacities depend on many factors and dynamics (Gallopin, 2006), including socio-demographic and economic factors such as education, age, and income; institutional factors such as prevention messages, access to clinics, and trained medical staff; and technological factors such as access to media and prevention tools and geospatial analysis capabilities (Birkmann, 2006). Socio-demographic data that could help identify the vulnerability of human populations in spatial terms include elements such as: population distribution, density, age, and gender; education and income; and specific public health-related information that often varies from country to country. In some studies of vector-borne diseases that are transmitted human to human (as in the example of malaria given below) the relationships of risk, exposure, susceptibility, and resilience are somewhat changed so that *risk* comprises the mostly entomological component (exposure) as well as the human population components of susceptibility and resilience combined together as *vulnerability* (Haines *et al.*, 2006).

Examples of recent research

Since vulnerability of human populations to vector-borne disease is influenced by complex biophysical, social, and human behavioral factors, the assessment of its spatial and temporal dynamics and associated environmental and anthropogenic patterns requires sophisticated mapping and modeling techniques.

As an example of empirical studies that offer options for mapping vulnerable populations, Cleckner and Allen (2014) demonstrated that a dasymetric mapping technique could be used successfully to map spatial patterns of vulnerable human populations to mosquito vector exposure. They refined vulnerable population data available at the level of census districts to a finer spatial scale by using satellite imagery data, thus narrowing down the areas where relief programs needed to be focused. In the same vein, Gadiaga *et al.* (2021) have developed an integrated housing quality-based typology of the neighborhoods in Dakar, Senegal, by combining the 2013 census data with remotely sensed land cover and land use data at a very high resolution. The derived housing quality indices appear to be relevant proxies of spatial variations of crude mortality rate and could therefore be considered as a guide for public health interventions in cities where accurate and detailed health data remain limited. Van Wesenbeeck *et al.* (2016) linked susceptibility survey data for human populations collected at the household level with agro-ecological data obtained from satellite image analysis at a large spatial scale for regions in East and West Africa. The satellite data analysis aimed at locating areas where ecosystem sensitivity to the effects of climate change was the highest. Bantis *et al.* (2017) mapped the spatio-temporal patterns of disabled people during a major storm event in the UK in 2013. They used the data of automatic transportation fare collection to track the variation in time and space of the human population that was using London's transport network. This information included age categories of people using the network, thus allowing for the determination of movement – or lack thereof – of users more susceptible during emergency situations, such as the elderly, disabled people, and children.

Many empirical studies exist on the relevance of using EO data in characterizing populations and places most vulnerable to health risks (Weng *et al.*, 2014). Recently, Parselia *et al.* (2019) published a scoping review of the use of EO data in epidemiological modeling of malaria, dengue, and West Nile virus. The review shows that EO data are rarely combined with demographic data and, when they are, it is often only to consider population density. However, epidemiological models gain in accuracy if they integrate information on the characteristics of the populations exposed to the entomological threat. It is the vulnerability of these populations to this threat that determines the risk level of a disease outbreak.

Challenges and questions

The spatio-temporal representation of human vulnerability is a challenging task. At a technical level, the challenge is related to appropriately matching spatial detail of the different information sources during risk map integration exercises, as mismatches can cause inaccuracies and sometimes loss of spatial cohesion. At a conceptual level and in terms of data sourcing, the challenge lies in selecting and skillfully utilizing EO data to supplement or refine socio-demographic information that could be provided by census surveys or relevant individual and household surveys. Pertinent questions that address these challenges are:

- To what extent are EO and geospatial data suitable for developing the relationship with physical-environmental features and human factors in addressing vulnerability issues? Which EO and other geospatial data sources can be advantageous for mapping or monitoring conditions, patterns, or dynamics of susceptible population?
- How can one improve the integration of different geospatial data layers extracted from EO or other sources to capture the spatial-temporal variation of human vulnerability?

Responses and options

Below are key comments and suggestions from experts regarding critical geospatial data on vulnerable populations:

- In general, EO can provide baseline information on the location of the population, including identifying rural and urban areas; detailed and up-to-date EO data can provide evidence of hard-to-reach "invisible" populations, such as those that are dislocated and/or migratory.
- EO can provide frequent and detailed data to determine habitable vs. non-habitable areas in disaster situations for a variety of environments.
- Risk and sensitivity maps of communities and populations should incorporate public health-related elements that contextualize census data, EO data in combination with 3D settlement data, and vegetation data (if relevant to the disease).
- Multi-temporal EO analyses and change detection methods should be used.

From *Anopheles* to humans: reconstructing the risk of malaria infection in Dakar, Senegal

In this example, the definitions of risk and vulnerability are those for a human-to-human transmitted vector-borne disease, i.e. risk comprises entomological hazard plus vulnerability as the human population aspects of susceptibility and coping capacity. For malaria infection to occur, three components need to interact: the parasite, the vector, and the human host. The identification of areas where these three components can easily interact is essential in the fight against malaria and the improvement of programs for the prevention and control actions and to guide interventions toward controlling the disease.

Studying the risk of malaria infection in urban spaces requires detailed and high-quality information on the presence of vector-competent mosquitoes, on the individual behaviors of the human hosts, and on the parasites. The provision and utilization of such information comes at a significant cost. In resource-poor countries, researchers and public health practitioners often have limited resources and often inadequate data on prevalence and incidence of the disease, including poor representation of the actual population affected (Programme National de Lutte contre le Paludisme, 2008; Diallo *et al.*, 2012). In addition, relevant geospatial data are limited and often non-existent.

With the development and more widespread application of GIS and satellite imagery, and more diversified sources of ecosystem-related geospatial information ecosystems, it is feasible to extract key environmental variables related to mosquitoes and their breeding sites. The combination of such information with sociodemographic census data or health surveys enables researchers to reconstruct and study the spatial variability of malaria infection risk (Borderon, 2016).

For the city of Dakar, Senegal, multi-temporal satellite imagery, census data, and results from social and health surveys have been integrated into a GIS with the goal of identifying potential exposure of urban neighborhoods and populations to epidemic risk of malaria (Fig. 2.5.1). Epidemic risk has been defined as the combination of two key indicators of malaria infection: the presence of the *Anopheles* spp. vectors, and social vulnerability of individuals or populations regarding the exposure to these vectors. Prevalence data collected during Project ACTUPALU were used to validate the risk model and produce a risk map (Borderon and Oliveau, 2017).

Expected outcomes and impacts

What does this map do? The risk map associates each district of the city with a profile of exposure to the disease, highlighting areas of potential outbreaks. At an urban scale, the mapping results contribute to the identification of areas of social-ecological vulnerability and reveal a possible risk pattern of malaria transmission for different types of sub-urban areas.

This risk model was created from the combination of three indicators:

1. An estimate of the human biting rate (HBR) of mosquitoes.
2. An estimate of the precise population density and dilution effect on biting rates.
3. An estimate of the social vulnerability of the population.

These indicators have been produced and aggregated at the Census District (CD) level according to a conceptual vulnerability and risk framework adapted from Taubenböck *et al.* (2008) (Fig. 2.5.2). The CD represents the smallest administrative unit of approximately 1000 inhabitants within the metropolitan area of Dakar.

The risk concept derives from two parameters: hazard and vulnerability. In the case of this study, the hazard is the HBR (bites/person/night), representing the probability of being bitten by a mosquito vector. The notion of exposure – often implicitly related to the studied hazard – has been added by taking into account the population density: the higher the density, the more the effects of bites are theoretically shared among the population. The combination of these two parameters gives an approximation of the probability of being bitten, all things being equal. However, the probability of individuals or populations being bitten can vary dramatically according to education, resources, or demographic characteristics, among other factors.

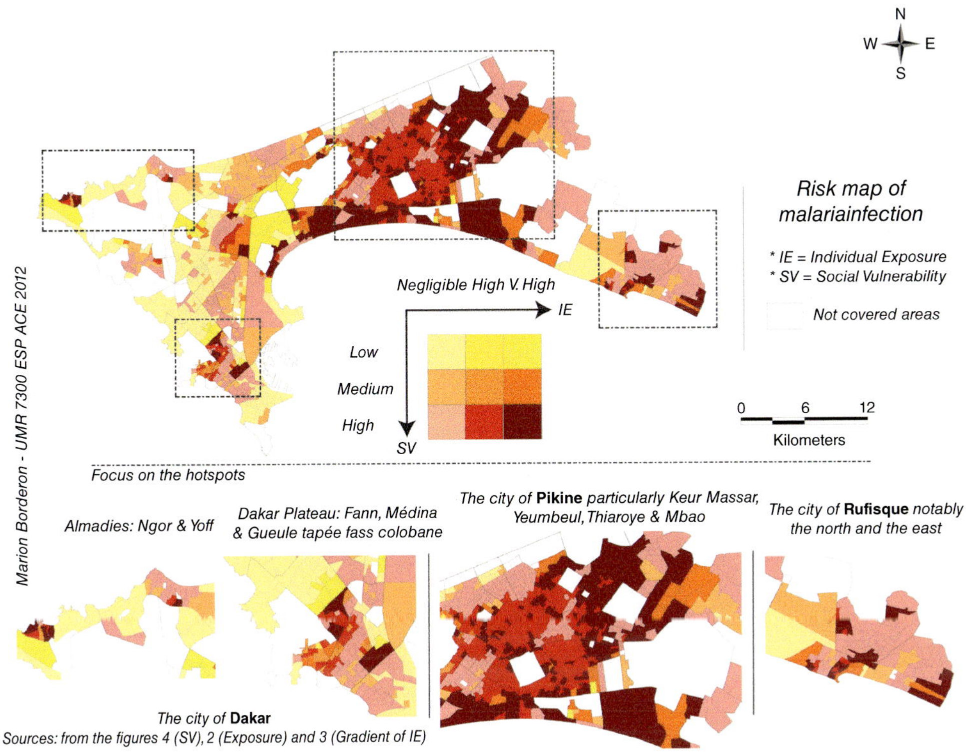

Fig. 2.5.1. Empirical risk model of malaria infection in Grand Dakar, Senegal.

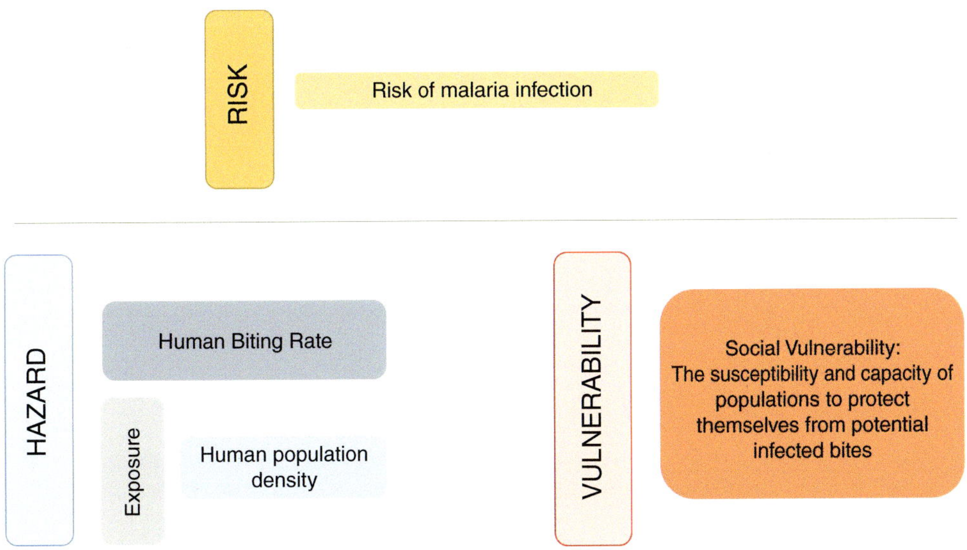

Fig. 2.5.2. Conceptual framework applied to the risk of malaria infection. (From: Taubenböck *et al*., 2008.)

In the multivariate map (Fig. 2.5.1), risk is represented as a linear combination or aggregation of hazard and vulnerability. Nine different combinations of risk have been developed. Individual exposure (IE) to mosquitoes is divided into three categories: negligible, high, and very high. The social vulnerability ranking also entails three categories (low, medium, and high); it assumes that the higher the social vulnerability is, the lower the protection against the bites will be and the lower the likelihood of available health care service will be.

The color intensity scheme of the map points to the potential hotspots of malaria infection, thus identifying areas that deserve special attention during the rainy season, especially if there is little or no medical support for households by institutions or aid programs.

Two weak points deserve to be considered regarding this mapping approach. One involves the selection of thresholds for continuous variables and assumptions on mobility. The scientific literature on malaria infections contains little knowledge on the existence of thresholds whereby the probability of being bitten varies as a function of population density. The thresholds used are thus arbitrary detection thresholds. The second point involves the initial assumption that the hosts are bitten where they live and that the model is static, representing a general situation. This means two things: the mobility of people and periods spent outside the area are not considered, and the effects of seasonal/environmental conditions are not taken into account.

Why use EO data? When working on diseases such as malaria, researchers often face a lack of quality data required for optimal targeting of the intervention and monitoring (Ceccato *et al.*, 2017; Quattrochi *et al.*, 2017). Data from satellite imagery with useful spatial and temporal resolution can help fill the gap and are becoming more readily available. EO data can be particularly helpful for characterizing relevant landscape features or urban environments. Although socio-economic or demographic data are rarely derived directly from EO, they can be combined or extrapolated to produce useful information, such as the malaria infection risk index in the above-mentioned work.

Who are the end users? Ideally, malaria risk maps and their subcomponents – hazard and vulnerability maps – become valuable tools for practitioners and policy makers who wish to obtain useful information on the potential hotspots of risk of malaria infection in urban environments. The information would allow them to identify vulnerable populations and address their needs, identify uneven capacity for preparedness and response, and reduce pre-existing risk. In the context of vector-borne diseases in low- and middle-income countries (LMICs), this knowledge, combined with geospatial information products developed with the help of remotely sensed data and GIS, can enable decision makers to better allocate limited resources in the fight against epidemics. Since vector-borne diseases are linked to climate and environmental conditions as well as human and societal characteristics, the integration of climate data and environmental information combined with socio-demographic data sets becomes an essential task for governmental and intergovernmental institutions with responsibilities for public health.

Technical considerations for producing risk and vulnerability maps

This section highlights the technical considerations necessary for the production of the different subcomponents of the risk map.

What EO data are needed? Assessing the risk of malaria infection requires a combination of data and skills and relies on interdisciplinary studies. Core aspects of the work are illustrated in Fig. 2.5.3. The integrated assessment of risk of malaria infection in Dakar required several sources of quantitative information, as listed in Table 2.5.1. A more detailed description of EO data used for the HBR model is given below. EO data and derived information products are integrated into hazard modeling of HBR estimates and identification and characterization of the human vulnerability to malaria infection.

EO data for estimating the HBR

A tele-epidemiology approach was used to estimate the density of the main mosquito vector of malaria in Dakar, *Anopheles gambiae* sl. The approach involved in the production of the vulnerability map depicted in Fig. 2.5.4 followed three major steps: (i) intensive ground measurements

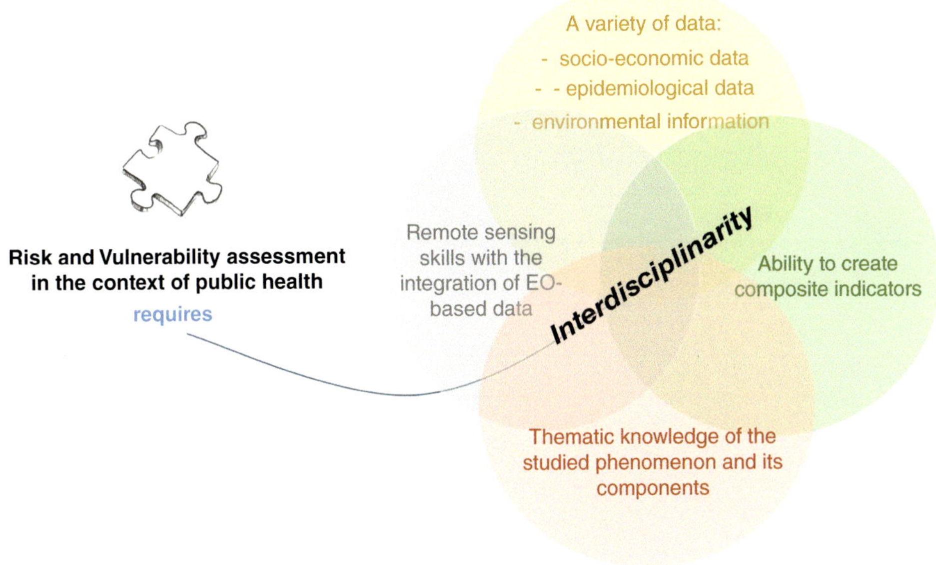

Fig. 2.5.3. The inherent interdisciplinarity of assessing risk and vulnerability.

Table 2.5.1. Preliminary data sources for Dakar metropolitan area and census districts (CDs).

Type	Area coverage	Time frame	Source
Socio-economic and demographic variables	2000 CDs	2002 (published in 2006)	Census ANSD
Predicted HBR	1476 CDs	1994–1997–2008–2010	Laboratoire d'aérologie
Prevalence (data used for risk model validation)	112 CDs	2008	ANR ActuPalu
Multi-temporal land use and land cover analysis	Dakar metropolitan area	1988–2008	Centre de Suivi Ecologique (CSE)

ANR, Agence Nationale de la Recherche; ANSD, Agence Nationale de Statistique et de la Démographie.

(*Anopheles* larval habitats and HBR); (ii) selection of satellite data for mapping and extracting environmental and meteorological information; and (iii) use of statistical models taking into account the spatio-temporal variability of the data.

The models were developed by a team of researchers at the Department of Infectiologie de Terrain de l'Institut de Recherche Biomédicale des Armées (IRBA) in Marseille (Machault *et al.*, 2012).

High-resolution SPOT-5 satellite images of Dakar and surroundings were acquired for the summer rainy season to coincide with the fieldwork during 26 September 2007, 24 September 2008, and 28 September 2009; a dry season image was captured on 11 May 2009. This multi-temporal, atmospherically corrected data set included three spectral bands at 2.5-m spatial resolution (green, red, and near infrared) and one short-wave infrared band at10-m spatial resolution (Machault *et al.*, 2012). A digital elevation model (DEM) at a spatial resolution of 90 m was available from the Shuttle Radar Topography Mission (SRTM 4.1).

In order to characterize the hazard component for a better HBR estimate one needs to consider the net population density in the built-up areas of a city. In Dakar, population densities are often calculated on the basis of CD data. However, since CDs are not completely covered by built-up areas, a more realistic measure can be applied in the form of dasymetric mapping. The principle of

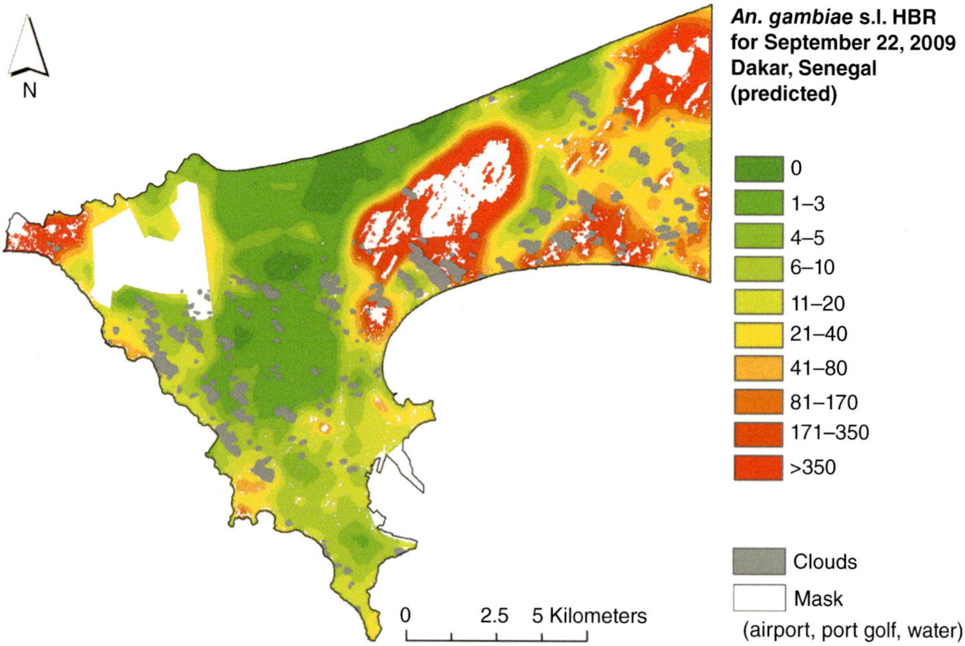

Fig. 2.5.4. Vulnerability of populations in terms of predicted *Anopheles gambiae* sl. number of bites per person per night for 22 September 2009. (From: Machault *et al*., 2012.)

dasymetric mapping is to adjust human population density exclusively to the space where people actually live (Mennis, 2003). Dasymetric mapping recalculates the actual – or net – population density by excluding areas of vegetation, water, bare soil, and roads. Figure 2.5.5 shows the urban net density for Dakar (Borderon *et al.*, 2014).

EO data for estimating human vulnerability characteristics: The actual risk map (Fig. 2.5.1) reflects the interaction between hazard and vulnerability of the population. In the context of this study, social vulnerability has generally been defined as the set of characteristics of a group or individuals in terms of their capacity to anticipate, cope with, and resist the impact of natural hazards like malaria infection. Following the Social Vulnerability Index approach,[1] a social vulnerability metric was implemented for mapping in the metropolitan area of Dakar (Borderon, 2016; Fig. 2.5.6).

What resources are needed? Assessing complex environmental and social phenomena associated with the geospatial assessment of malaria infection risk requires the integration of EO data with demographic, socio-economic, and other data. It is recognized that the exposure to hazard alone is insufficient to predict impact, because the populations affected are heterogeneous and vulnerable to impact in different ways (De Sherbinin, 2017). Common issues often identified in the literature with respect to data integration revolve around data quality and scale.

Data availability and the choice of scale: Since vulnerability assessments rely on a variety of data sources, reliable access to these data sets is essential and policies for data sharing embedded in a spatial data infrastructure are required to provide reliable and consistent results. Once the database is built, one of the challenging issues is the choice of a common scale for data integration. This choice should ideally be related to the "scale of action" at which phenomena or features of interest can best be observed, rather than by the scale of available data. For instance, small-scale data sets can be resampled at higher resolution. Fritzsche *et al.* (2014) and OECD (2008) discussed different aggregation methods in detail. Data integration also requires familiarity with the science behind the data sets and their composition. As an example, many remote sensing data analyses involve highly refined methods for measuring and assessing the impact

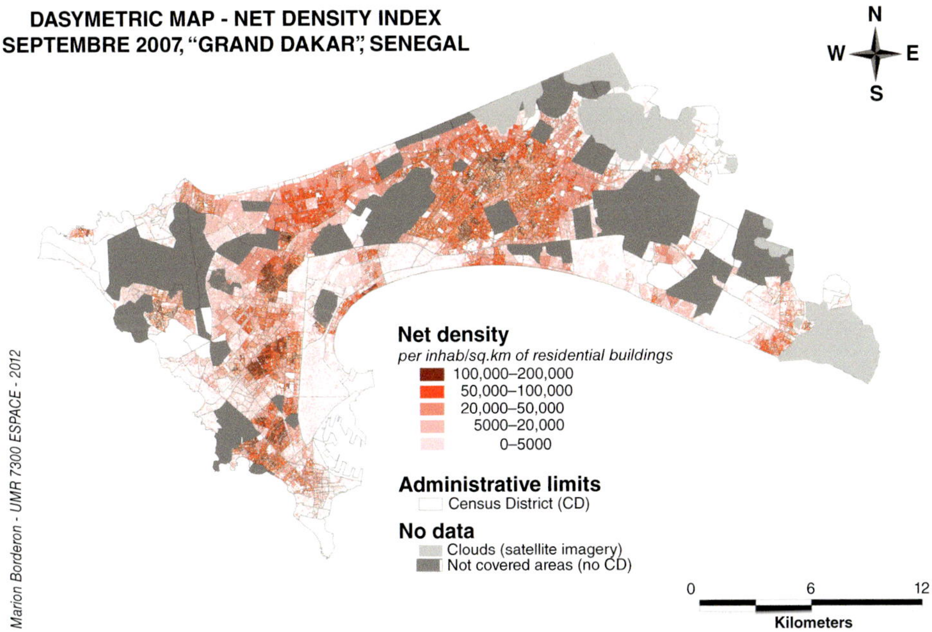

Source: Image SPOT 5 of 2007 during the rainy season – 2.5m resolution – 10m SWIR – No supervised Isodata classification - corrected. Breakdown of population densities by CDs on areas of residential buildings. Laboratoire d'aerologie: V. Machault, C. Vignolles and JP. Lacaux

Fig. 2.5.5. The estimation of urban net density in Dakar.

of errors in their measurements through classification accuracies and standard errors, whereas most spatial socio-economic data do not come with corresponding error bars for the estimates contained in them. Hence, characterizing the validity and accuracy of derived products can be challenging (De Sherbinin, 2017).

Data management and analysis: Recently, computer processing, data storage facilities, and access to remotely sensed products have become more ubiquitous and user-friendly. Moderate- and high-resolution satellite imagery is often available free of charge. However, there is still a significant need for assistance in the process of technology transfer. This applies especially for the public health sectors of many countries that seek to employ effective geospatial assets against the threat of malaria with the help of EO data analysis.

Perspectives

What future developments are needed? Integrative vulnerability assessments require some methodological advances. The difficulty of coupling the dynamic disease modeling approaches and their uncertainty to assess population vulnerability is often evident in projects modeling the risk of malaria infection and transmission. More transparency is needed at each stage of the decision process when it comes to combining modeling and visualizing the data. Systematic validation of the produced model is another requirement. In the context of malaria, this could be achieved, in part, by the development of new frameworks for modeling routine malaria surveillance data and by the integration of other malaria-related metrics.

What are the opportunities for EO? EO data, products derived from them, and other geospatial data from available databases and surveys offer several opportunities for the development of risk and vulnerability maps with high spatial and temporal resolution:

- *To better understand the relationship between diseases and the environment and climate.* Satellites provide raw data that are continuously archived and cover large areas

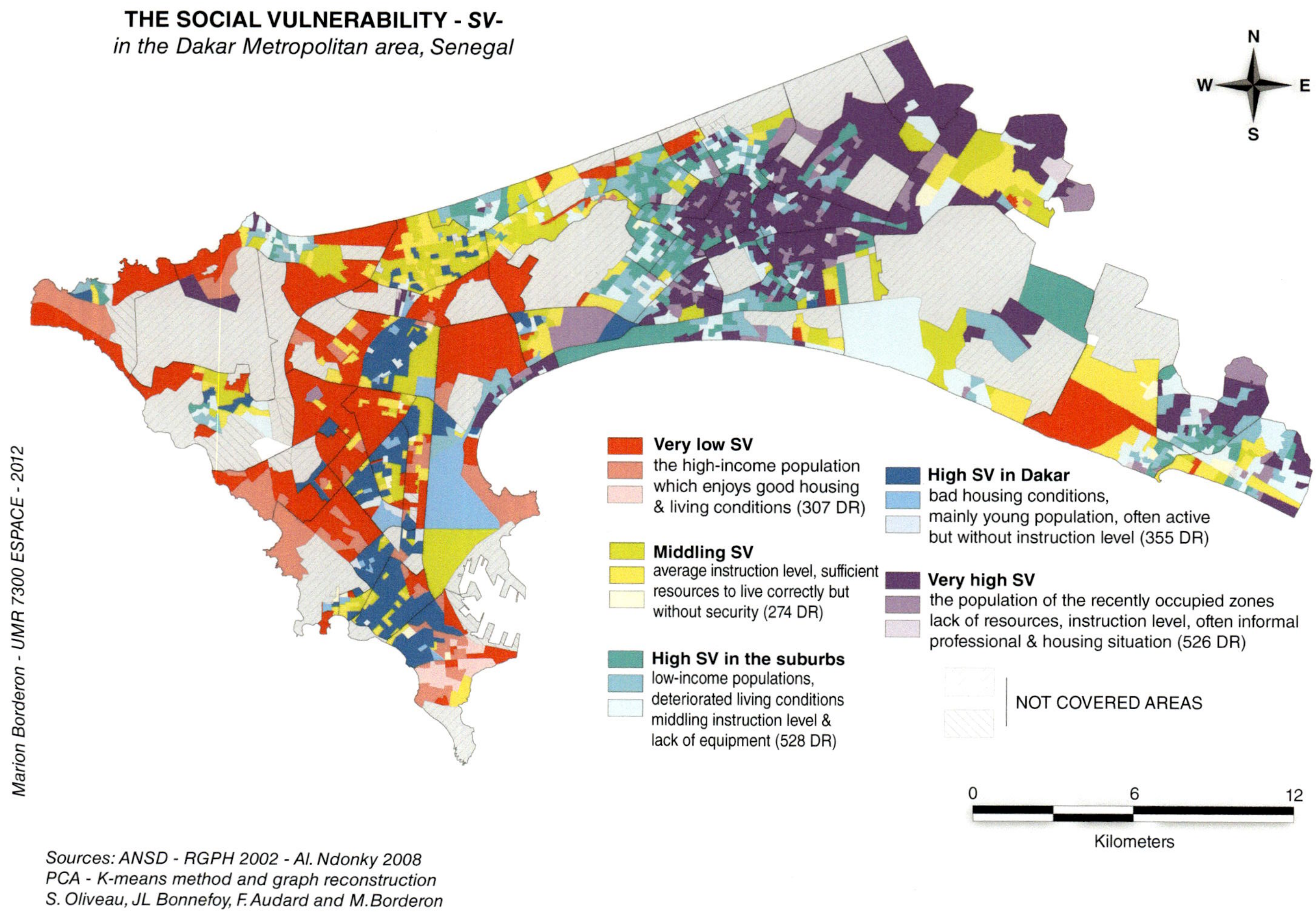

Fig. 2.5.6. The Social Vulnerability Index for Dakar, Senegal.

of the Earth. Sources for satellite-generated climate or environmental data that can help assess the exposure of population to a climate-related disease are varied and some are freely available online. A list of useful sources is included in Quattroci *et al.* (2017).

- *To assess precisely who is at risk and where.* Population data sets from surveys or census data can be combined with EO data to produce dasymetric maps. Dasymetric mapping is a method to disaggregate census data to finer scales by integrating satellite-based data and land cover data; the result provides a more realistic impression about the population distribution than using arbitrary administrative boundaries such as census tracts.
- *To better understand the relationship between diseases and demographic and socio-economic characteristics of the population.* Some programs combine EO data with population census data and surveys to provide high-resolution multi-temporal population maps. These maps offer estimates of population size and distribution as well as other related characteristics for data-limited environments. For instance, the *WorldPop* population mapping program is based on peer-reviewed methodologies and currently provides the spatial demographic data sets of choice for over 100 government agencies in low-income and lower-middle-income countries in Africa and Asia.
- *To help measure the accessibility of health care or the health situation of some marginalized populations* (not present in the classical data sets) or a population in a post-crisis situation where key baseline data are not available. Satellite-based information could help to establish baseline data for population surveys in the absence of household lists or systematic civil registration. EO products can support selection of representative samples of populations in the absence of census-type data on households (Kondo *et al.*, 2014).

Current EO product developments in the public health sector: The integrative assessment of risk and vulnerability in the context of public health has been applied and improved over the past 30 years. Some examples of new products and approaches utilizing EO in the public health sectors of vulnerable countries reflect well on current developments.

Example of bottom-up population mapping from WorldPop

Where census data are outdated or unreliable, *WorldPop* has been collaborating with the Bill and Melinda Gates Foundation and Oak Ridge National Laboratories to develop approaches to estimating population distributions at high spatial resolution through a combination of satellite-derived feature extractions and household surveys. Initial outputs are available on the *WorldPop* website with some outputs already available for Nigeria in their vaccination tracking system.[2]

The *Flowminder Foundation* offers examples of the collection, aggregation, integration, and analysis of anonymous mobile operator, satellite, and household survey data.[3] The following is an extract from the three applications of their work related to public health:

Disaster response	"We pioneered the use of de-identified data from mobile operators to follow population displacement. With this data we support relief agencies in delivering the right supplies to the right people at the right time."
Socio-economic analysis	"Traditional surveys in low- and middle-income countries produce estimates only for large areas. Using new statistical methods, satellite and mobile data we produce estimates of poverty and key social indicators at a resolution of 1km^2."
Precision epidemiology	"Most infectious diseases spread through human movements. We integrate large numbers of data sources, including data from mobile phone operators to model and predict spread of infectious diseases."

Notes

[1] http://artsandsciences.sc.edu/geog/hvri/sovi (accessed 4 January 2022).
[2] https://www.worldpop.org/methods (accessed 4 January 2022).
[3] http://www.flowminder.org/ (accessed 4 January 2022).

References

Bantis, T., Haworth, J., Holloway, C., and Twigg, J. (2017) Mapping spatiotemporal patterns of disabled people: The case of the St. Jude's storm emergency. In: Griffith, D., Chun, Y., and Dean, D. (eds.) *Advances in Geocomputation. Advances in Geographic Information Science*. Springer, Cham, Switzerland. https://doi.org/10.1007/978-3-319-22786-3_10

Birkmann, J. (2006) Measuring vulnerability to promote disaster-resilient societies: Conceptual frameworks and definitions. In: Bemmerlein-Lux, F. (ed.) *Measuring Vulnerability to Natural Hazards: Towards Disaster Resilient Societies*. Deutsche Gesellschaft für internationale Zusammenarbeit GmbH (GIZ), Germany, pp. 9–54.

Blaikie, P., Cannon T., Davis I., and Wisner, B. (1994) *At Risk: Natural Hazards, People's Vulnerability, and Disasters*. Routledge, London.

Borderon, M. (2016) Entre distance géographique et distance sociale: le risque de paludisme-infection en milieu urbain africain: l'exemple de l'agglomération de Dakar, Sénégal. Doctoral dissertation, Aix-Marseille.

Borderon, M. and Oliveau, S. (2017) Vulnérabilités sociales et changement d'échelle. L'espace compte! *Space Populations Societies.* DOI: https://doi.org/10.4000/eps.7012.

Borderon, M., Oliveau, S., Machault, V., Vignolles, C., Lacaux, J.P. *et al.* (2014) Qualifier les espaces urbains à Dakar, Sénégal. Résultats préliminaires de l'approche croisée entre télédétection et données censitaires spatialisées. *Cybergeo: European Journal of Geography* 670. DOI: https://doi.org/10.4000/cybergeo.26250.

Ceccato, P., Connor, S., Dinku, T., Kruczkiewicz, A., Lessel, J. *et al.* (2017) Integrating remotely sensed climate and environmental information into public health. In: Dale, A., Quattrochi, E.W., Wentz, E., Lam, N.S.-N., and Emerson, C.W. (eds) *Integrating Scale in Remote Sensing and GIS*. Routledge, New York, pp. 303–335.

Cleckner, H. and Allen, T.R. (2014) Dasymetric mapping and spatial modeling of mosquito vector exposure, Chesapeake, Virginia, USA. *ISPRS International Journal of Geo-Information* 3, 891–913. DOI: https://doi.org/10.3390/ijgi3030891.

Cutter S.L. (1996) Vulnerability to environmental hazards. *Progress in Human Geography* 20, 529–539. DOI: https://doi.org/10.1177/030913259602000407.

De Sherbinin, A. (2017) Remote sensing and socioeconomic data integration: Lessons from the NASA Socioeconomic Data and Applications Center. In: Quattrochi D.A., Wentz E., Lam N.S.-N., and Emerson C.W. (eds) *Integrating Scale in Remote Sensing and GIS*. Routledge, New York, pp. 371–388.

Diallo, A., Ndam, N.T., Moussiliou, A., Dos Santos, S., Ndonky, A. *et al.* (2012) Asymptomatic carriage of *Plasmodium* in urban Dakar: The risk of malaria should not be underestimated. *PLoS ONE* 7. DOI: https://doi.org/10.1016/j.gloenvcha.2010.07.003.

Fritzsche, K., Schneiderbauer, S., Bubeck, P., Kienberger, S., Buth, M. *et al.* (2014) Vulnerability sourcebook: Concept and guidelines for standardised vulnerability assessments. *GIZ.* Available at: https://www.adaptationcommunity.net/download/va/vulnerability-guides-manuals-reports/vuln_source_2017_EN.pdf (accessed 14 June 2021).

Gadiaga, A.N., De Longueville, F., Georganos, S., Grippa, T., Dujardin, S. *et al.* (2021) Neighbourhood-level housing quality indices for health assessment in Dakar, Senegal. *Geospatial Health* 16. DOI: https://doi.org/10.4081/gh.2021.910.

Gallopin, G.C. (2006) Linkages between vulnerability, resilience, and adaptive capacity. *Global Environmental Change* 16, 293–303. DOI: https://doi.org/10.1016/j.gloenvcha.2006.02.004.

Haines, A., Kovats, R.S., Campbell-Lendrum, D., and Corvalán, C. (2006) Climate change and human health: Impacts, vulnerability and public health. *Public Health* 120, 585–596. DOI: https://doi.org/10.1016/j.puhe.2006.01.002.

Kondo, M.C., Bream, K.D., Barg, F.K., and Branas, C.C. (2014) A random spatial sampling method in a rural developing nation. *BMC Public Health* 14, 338. DOI: https://doi.org/10.1186/1471-2458-14-338.

Machault, V., Vignolles C., Pages F., Gadiaga L., Tourre M.Y. *et al.* (2012) Risk mapping of *Anopheles gambiae s.l.* densities using remotely-sensed environmental and meteorological data in an urban area: Dakar, Senegal. *PLoS ONE* 11. DOI: https://doi.org/10.1371/journal.pone.0050674.

Mennis, J. (2003) Generating surface models of population using dasymetric mapping. *The Professional Geographer* 55, 31–42.

Organisation for Economic Co-operation and Development (OECD) (2008) *Handbook on Constructing Composite Indicators: Methodology and User Guide*. OECD, Paris. Available at: https://www.oecd.org/sdd/42495745.pdf (accessed 15 June 2021).

Parselia, E., Kontoes, C., Tsouni, A., Hadjichristodoulou, C., Kioutsioukis, I. *et al.* (2019) Satellite earth observation data in epidemiological modeling of malaria, dengue and West Nile virus: A scoping review. *Remote Sensing*, 11, 1862. DOI: https://doi.org/10.3390/rs11161862.

Programme National de Lutte contre le Paludisme (PNLB) (2008) Morbidite et Mortalite Palustre au Sénégal en 2008: Synthèse donneés districts & hospitaux. Dakar, Sénégal.

Quattrochi, D.A., Wentz, E., Lam, N.S,-N., and Emerson, C.W. (eds) (2017) *Integrating Scale in Remote Sensing and GIS*. Routledge, New York.

Taubenböck, H., Post, J., Roth, A., Zosseder, K., Strunz, G. *et al.* (2008) A conceptual vulnerability and risk framework as outline to identify capabilities of remote sensing. *Natural Hazards and Earth System Science* 8, 409–420. DOI: https://doi.org/10.5194/nhess-8-409-2008.

Turner, B.L. (2010) Vulnerability and resilience: Coalescing or paralleling approaches for sustainability science? *Global Environmental Change* 20, 570–576. DOI: https://doi.org/10.1016/j.gloenvcha.2010.07.003.

Van Wesenbeeck, C.F.A., Sonneveld, B.G., and Voortman, R.L. (2016) Localization and charactorization of populations vulnerable to climate change: Two case studies in sub-Saharan Africa. *Applied Geography* 66, 81–91. DOI: https://doi.org/10.1016/j.apgeog.2015.11.001.

Weng, Q., Xu, B., Hu, X., and Liu, H. (2014) Use of earth observation data for applications in public health. *Geocarto International*, 29, 3–16. DOI: https://doi.org/10.1080/10106049.2013.838311.

2.6 EO and Geospatial Data Utilization During the COVID-19 Pandemic: A Preliminary Appraisal

Dirk Werle[1] and Guy Aubé[2]
[1]*Ærde Environmental Research, Halifax, Canada;* [2]*Canadian Space Agency, Saint-Hubert, Québec, Canada*

Context

Two decades into the third millennium, the 2019 coronavirus disease (COVID-19) pandemic is affecting the human population in ways that present generations have not experienced. Not since the Spanish flu a century ago has it been more apparent that a public health crisis in one community can quickly become a crisis around the world. Travel has never been faster or more frequent, trade more global, and human populations so numerous. Institutions have been warning of a potential pandemic for decades, citing climate change, rapid urbanization, and our increasing proximity to viral reservoirs like farm animals and wildlife (US Department of Homeland Security, 2006; GHRF Commission, 2016; WHO, 2005, 2016). Many scientists have posited that the source of the virus (SARS-CoV-2) was direct contact with wildlife like bats and pangolins, which led to the human-to-human transmission of COVID-19 (Andersen *et al.*, 2020; Zhang *et al.*, 2020). Piecing together what happened and how to address it has been the preoccupation of a great many people working in diverse fields. Whether doing fieldwork, working in laboratories, rolling out public health measures, compiling spatial data, or studying satellite images from hundreds of kilometers above the Earth's surface, scientists and practitioners are optimizing and capitalizing on each other's knowledge and expertise (Franch-Pardo *et al.*, 2020). They are forming new One Health research collaborations to more fully appreciate the complex web of natural and human-induced hazards that play roles in creating public health crises.

This preliminary appraisal of the utility of geospatial EO data in support of public health and safety measures was undertaken during the COVID-19 pandemic. Prior to the outbreak, EO research and development efforts geared toward the pandemic thematic have been scarce (Timpka *et al.*, 2011; Jonas, 2013); in-depth risk-related assessments and a dedicated EO-related pandemic playbook are not available. Hence, this appraisal is confined to EO-based information and knowledge generation as it pertains to current environmental conditions and changes, public health, and public safety and surveillance-related issues. Further attention is directed at the contribution of geospatial data in general and GIS technology in particular (Esri, 2011), as it has directly impacted and mediated situational reporting during the first half of 2020 (Franch-Pardo *et al.*, 2020). It is anticipated that comprehensive assessments and analytical evaluations of EO and GIS contributions in support of public health responses to the pandemic will be conducted in the post-COVID-19 era.

Collaboration and participation of the EO and geospatial community

The rapid global emergence of COVID-19 has made it obvious that the effects of the pandemic are pervasive and that working together is in everyone's best interest. The response has inspired a spirit of collaboration among scientists, managers, administrators, doctors, researchers, engineers, and the public at large on a scale rarely before seen. Members of the geospatial and EO communities have maintained essential data collection and curatorial services and provided evidence-based data, information, and know-how to support decision makers at various institutional and political levels. The pandemic has certainly laid bare the importance of openness of data, research outcomes, and research infrastructure (Kituyi, 2020), pointing toward the necessity of establishing best practices for pandemic strategies (Gibney, 2020) through synthesizing evidence (Donnelly *et al.*, 2018).

The all-hands-on-deck outlook has also been adopted by the geospatial community, which has been actively and creatively searching for ways to directly or indirectly contribute to the needs of public health. For example, the ESA held the Global Space and Economic Workshop in July 2020 to discuss the value of EO data during and after the COVID-19 crisis, how space data can help with monitoring the pandemic, and if it could even assist with recovery in the post-COVID world.[1] During the COVID-19 lockdown period, the Secretariat of the GEO initiated a series of virtual meetings. They encouraged participants from many different countries to share information and invited them to present their ideas on how best to place EO at the service of public health. They have been updating their website[2] with their initiatives and those of their members, participating organizations, and associates on how EO has been supporting response and recovery actions related to the COVID-19 pandemic.

Following an ESA call for rapid action, ESA, NASA, and Japan Aerospace Exploration Agency (JAXA) implemented an open access COVID-19 Earth Observation Dashboard[3] as a joint space agency initiative. It demonstrated the capability of satellite sensors and highlighted dashboard visualization technology. EO image capture relevant to the pandemic situation and measurements of environmental conditions documented numerous changes as countries closed businesses, people deserted public spaces and avoided travel, and industries wound down production. Ultimately, the EO dashboard initiative represented an important signal by the EO community and its leadership that there is a commitment to act and participate in the response and recovery process.

NASA devoted its 2020 Space App challenge[4] to exploring fresh perspectives on how EO data might help confront the COVID-19 pandemic. The goal was to document observable changes and combine EO environmental data with epidemiological and demographic data to create new knowledge. During the 48-h event, more than 1400 projects were created by over 15,000 participants and problem solvers from 150 countries. They came together to form more than 2000 teams. Experts on the various subject matters came from space agencies in the USA, Canada, Europe, and Japan, and were available online to answer questions and stimulate discussion on EO data fusion with socio-economic, environmental, demographic, and medical data.

Near-real-time mapping and monitoring

During the early phase of the COVID-19 pandemic, the collection, visualization, and analysis of geospatial data related to affected areas and near-real-time situational awareness played an important role in decision making and public relations. Historically, tracing and mapping disease outbreaks has deep roots, largely born out of the necessity of finding a way to keep people living in populous cities healthy. From the bubonic plague in the mid-14th century, to John Snow in the mid-19th century proving that cholera outbreaks in London were from contaminated communal wells, to the current day, disease mapping has provided us with vital information. Presently, big data and GIS, however, have brought the functionality and relevance of disease maps to a whole new level. There have been many websites and portals created by public, private, educational, and other institutions (with some redundancy) to share resources about the pandemic. The WHO's Situation Dashboard[5] provides current numbers of infected people, deaths, affected countries, and other data regarding the pandemic. GIS software providers also engaged in tracking COVID-19[6] have provided similar statistics and offered story maps on mapping the outbreak on a daily basis,[7] thus increasing accessibility to responders new to the sector and helping build engaged citizens.

Among the maps of disease spread created to date, none has been as popular as the dashboard built by Johns Hopkins University's (JHU) Center for Systems Science and Engineering. The developers used an online cloud-based mapping, analysis, and data storage system to create, share, and manage maps, image layers, apps, and other geographic content (Dong *et al.*, 2020). At its core is the automated collection of data from health officials around the world to generate a current situation map showing locations and numbers of cases of COVID-19. Their decision to make their map free to everyone and make their underlying data set available made it go viral; at its highest point the map was receiving more than a billion interactions a day by people visualizing the map and mining the underlying data.[8] The example set by the JHU-operated tool has since been emulated by numerous other institutions mandated to provide pandemic and health information to the public.

Thus far, near-real-time or archived EO data has not featured directly in these public health dashboards. However, GIS layers referring to meteorological and environmental information often rely on analyses derived from EO satellites. Through media reports on COVID-19, the scientific value of EO data and geographically linked information, together with the basic visual appeal of satellite imagery, appears to have increased the attention of the general public. This trend has in part been stipulated by initiatives of the EO, GIS, and public health community of practice.

Public health situational awareness and related surveillance

The pandemic has exponentially increased the demand for geographically correlated information in support of public health situational awareness and related surveillance activities. Situational awareness, as defined by the US Department of Health and Human Services in 2015, results from the process of active information gathering (both domestic and international) with appropriate analysis, integration, interpretation, validation, and sharing of information linked to health threats and the health of the human population. Related to situational awareness but not synonymous, surveillance is a key information-gathering activity that encompasses timely human disease surveillance, animal disease surveillance, environmental monitoring, and gathering of intelligence and other information for early warning.

While gaining situational awareness and relying on surveillance by way of satellite imagery has been a common practice in the military intelligence communities for decades, there have been far fewer efforts of similar scope with EO satellites in other sectors, including public health. The COVID-19 outbreak has altered this practice, with the same service providers offering to apply their high-resolution Earth Intelligence satellite data acquisition capacity and analysis capability to the pandemic efforts. This offer was sustained by numerous and much-publicized demonstrations of imaging capabilities (Figs 2.6.1–2.6.4).[9] Major satellite operators showed that EO data can be used to provide near-real-time situational awareness for authorities regarding the effect of public health, safety, and security measures enacted during the pandemic. Minetto *et al.* (2020) measured human and economic activities from satellite imagery to support city-scale decision making.

The capacity to map, analyze, assess, and project the state of real-world problems has expanded in the wake of the Ebola crisis in West Africa between 2013 and 2016. EO and GIS were used to update and create maps to decide where to allocate treatment centers, how many beds to put in them, how to help patients get to the centers, and how to manage safe burials; it was arguably a turning point in the deployment of satellite technology for managing epidemics (Peckham and Sinha, 2017).

During the early stages of the pandemic, a GIS industry leader produced a White Paper entitled Geographic Information Systems for Coronavirus Planning and Response. It offered a timely overview for leaders and decision makers explaining how GIS systems can provide location

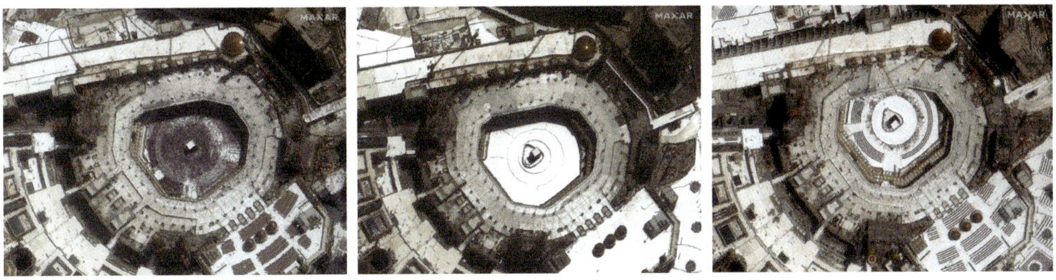

Fig. 2.6.1. High-resolution satellite imagery of public spaces showing density of human gatherings before and during pandemic-related restrictions exemplified by numbers of Muslim worshippers surrounding the Kaaba in Mecca, Saudi Arabia, on 14 February 2020 (left), 9 March 2020 (middle), and 3 April 2020 (right). In early March, Saudi Arabia took the rare step of limiting access to the Grand Mosque in Mecca, affecting the Umrah pilgrimage of many worshippers from around the world. (From: WorldView-3 satellite images ©2020 Maxar Technologies, used with permission.)

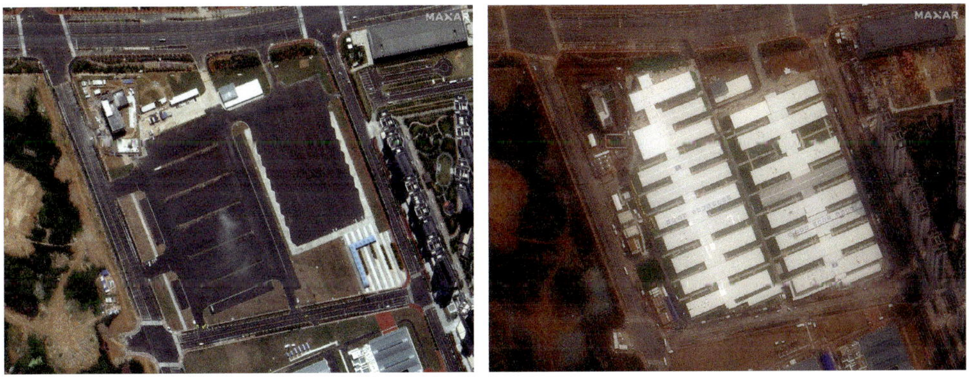

Fig. 2.6.2. High-resolution satellite imagery of construction activities at the Leishenshan hospital site in Wuhan, China on 3 August 2019 before the lockdown (left) and after completion on 4 March 2020 (right). A time series of the EO data provided evidence of immediate and significant efforts on the part of the authorities to increase medical care infrastructure in response to the COVID outbreak in that city. (From: WorldView-3 satellite images courtesy of Maxar Technologies, used with permission.)

intelligence (Esri, 2020). For public health authorities, geospatial information can help guide several activities, including making decisions about resource allocation, communicating more effectively with other agencies, and identifying vulnerable populations.

Potential EO applications

EO can contribute to assessing the risk of emerging diseases, modeling, and tracking outbreaks. During the COVID-19 outbreak, EO satellites continued their routine operation, as they are considered an essential service. Their data collections over large areas allowed analysts to quantify change at various spatial and temporal scales over selected sites or regions, sometimes producing stunning visuals. EO has the capacity to support pandemic response in two different ways: for near-real-time situational awareness and up-to-date reference imagery; and as analysis-ready data for ongoing risk mapping and modeling. Potential applications include the following: assessing population mobility, monitoring of air quality, analyzing weather conditions, monitoring social-economic activities, and utilizing up-to-date land use and land cover information, in conjunction with mobility and socio-economic data, for risk area identification.

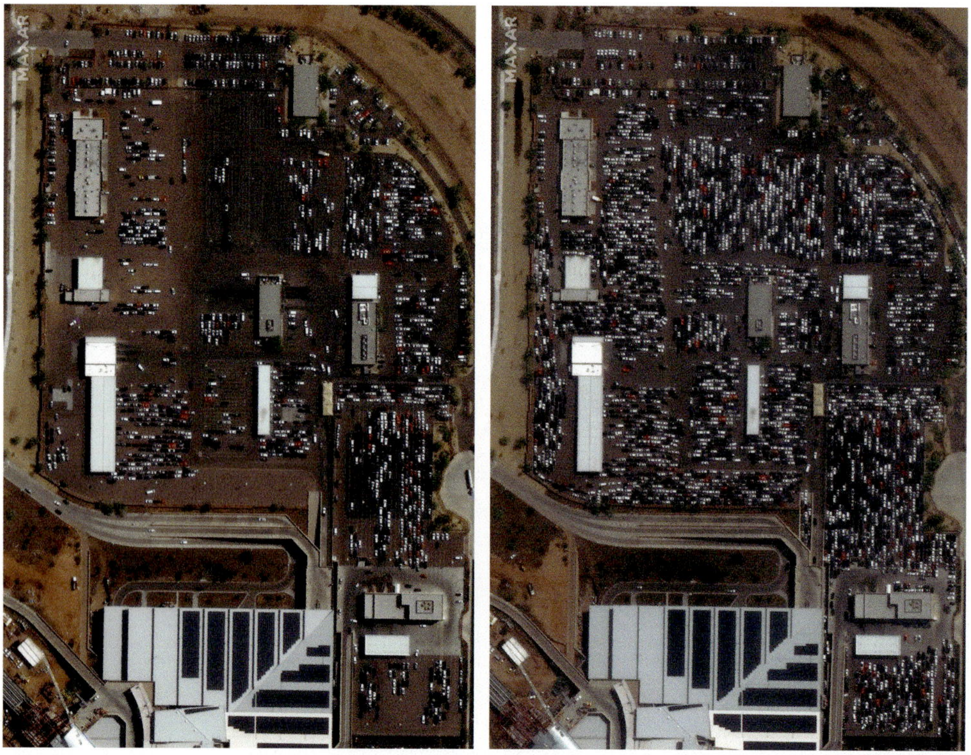

Fig. 2.6.3. High-resolution satellite imagery of the Phoenix Airport rental car center on 5 March 2020 (left) and again on 16 March 2020 (right) as fewer cars were being rented given increasing flight cancellations. (From: GeoEye satellite images ©2020 Maxar Technologies, used with permission.)

Weather

Although some respiratory viruses have distinct seasonal rhythms, whether there are seasonal factors effecting COVID-19 spread has yet to be determined definitively. Currently, many environmental variables such as rainfall and temperature are being examined, and there is, for example, some evidence that the virus might not be able to transmit as efficiently within environments with higher humidity (National Academies of Sciences, 2020; Ma *et al.*, 2020, 2021; Kifer *et al.*, 2021). However, the international virtual symposium on Climatological, Meteorological and Environmental (CME)[10] factors hosted by the World Meteorological Organization in August of 2020, made note of two observations about the COVID-19 pandemic. First, peer-reviewed published studies of SARS-CoV-2 transmission, COVID-19 incidence, morbidity, mortality, and recovery rates have failed to unequivocally show a robust and consistent response to temperature, humidity, wind, solar radiation, or other proposed meteorological drivers. This does not exclude the possibility of an influence, now or later in the pandemic, but the evidence presented is not sufficiently strong and the prevailing degree of uncertainty in existing studies renders the evidence unusable for further predictive applications at this stage. More research is needed to better understand which variables are most critical for dynamical understanding and COVID-19 risk prediction. Second, there is stronger evidence that air pollution in the form of fine PM affects the severity of COVID-19 symptoms. This is supported by evidence from both COVID-19 and other respiratory illness research that studies the impacts that chronic and acute $PM_{2.5}$ exposure can have on symptom severity. Because satellites produce a steady stream of weather-related data on a global scale, EO is well positioned to contribute to this research and forecasting of infection increase by providing data on air quality and on weather if and when these prove to be useful in predicting COVID-19 transmission.

Fig. 2.6.4. High-resolution satellite image of mass-grounding of commercial aircraft at Panama City International Airport, 24 March 2020. This pictorial representation of travel bans during the COVID-19 pandemic was a scene that repeated itself at many airports around the world. (From: WorldView-1 satellite image courtesy of Maxar Technologies, used with permission.)

Mobility

A novel feature of EO is its ability to pinpoint and visualize changes in levels of light at night. Nighttime radiance can be used to track urbanization and socio-economic parameters (population, GDP, etc.), evaluate disasters and armed conflicts, assess greenhouse gas emissions and energy use, and analyze light pollution and its effects on health. The global pandemic has offered scientists a unique opportunity to observe the sudden, mass changes in human behavior through the lens of satellites, which can provide near-real-time situational awareness, indicate the socio-economic burden of the disease, and monitor when normal activity has resumed.

In the USA, researchers found that nighttime radiance declined in states with lockdown measures but did not substantially decline in the states that did not enforce stay-at-home orders, such as Florida and Arizona (Elvidge *et al.*, 2020). This emphasizes that unless orders come from governing bodies, physical distancing will not occur on any significant level and high infection rates, like in Florida and Arizona, could be the result. Observations were made on other locations too: near Wuhan, where the virus first emerged, data showed that nighttime lights were brighter in residential areas around the city; dimmer nighttime lights were recorded in commercial inner-city areas during the lockdown (Fig. 2.6.5), suggesting that many people were

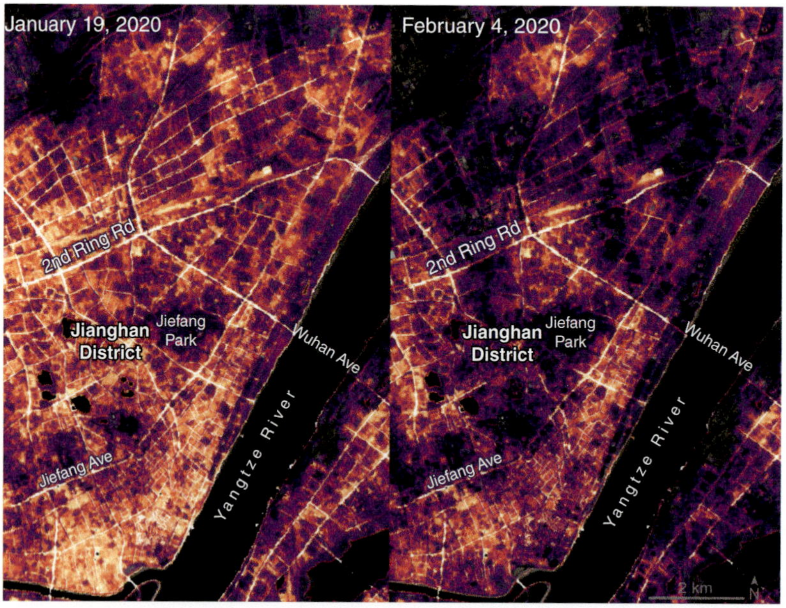

Fig. 2.6.5. Nighttime light conditions within the city of Wuhan, China, based on imagery collected on 19 January 2020 and 4 February 2020 by the VIIRS on the NOAA–NASA Suomi NPP satellite. Significant lighting changes are noticeable in the commercial Jianghan District and nearby residential areas, likely attributable to activity restrictions invoked by local authorities during late January in response to the COVID-19 outbreak. (From: NASA Earth Observatory/NASA Goddard Space Flight Center, VIIRS day–night band data from the Suomi National Polar-orbiting Partnership.)

staying home as requested (Liu *et al.*, 2020); India was substantially dimmer; London had dimmed although many nearby cities did not; Paris had dimming near its core; dimming was patchy in Italy and Japan; and Brazil and Iraq largely lacked any dimming (Elvidge *et al.*, 2020). Ultimately, these data can do more than help track mobility, compliance, and impact; it can give planners an indication of where (or whether) people might move, helping them prepare for subsequent lockdowns.

Mobile device data

Given the enormous accumulation of mobility data by telecommunication providers, disease mappers are able to access global databases that can be combined with information like disease incidence, bat or insect reservoirs, and even the travel of infected individuals. Mobility data can enhance the capacity of GIS to elucidate transmission dynamics and to coordinate and direct response to outbreaks. This is especially true in more remote areas and jurisdictions lacking electronic health records with up-to-date addresses. Technological GIS and telecommunications advances paired with the experience of fighting outbreaks of Ebola and other infectious diseases like SARS, H5N1, and H1N1 have helped officials not only locate disease but communicate with the general public, target services, and conceive of new ways to collect information that could directly contribute to expedient responses to outbreaks.

In Asia and Europe, contact tracing apps for mobile phones have been tested, in part certified, and made available, increasing our collective ability to respond to an outbreak. Accurate information on population movements is valuable for monitoring the progression of the outbreak and predicting its future spread, facilitating the prioritization of interventions and designing surveillance and containment strategies. The emergence of functioning contact tracing apps within a few weeks of conceptualization and testing is a vivid example of how rapidly geospatial and communication technologies can be paired, harnessed, and adapted to perform a critical task (Ferretti *et al.*, 2020), although there are considerable issues regarding privacy in some countries, which may limit the use of mobile device data (Lapolla and Lee, 2020).

Air quality

There has been much published on the link between human health and air quality with plenty of evidence that air pollution can both exacerbate and cause respiratory and heart diseases. EO makes short work of assessing air quality, providing details on near-surface concentrations of NO_2, SO_2, O_3, NH_3, volatile organic compounds (VOCs), and aerosol properties. NASA is presently developing the MAIA mission in partnership with epidemiologists and health organizations to use EO data to further study specific human health issues. MAIA will make radiometric and polarimetric measurements to characterize the sizes, compositions, and quantities of PM in air pollution; researchers will combine MAIA measurements with population health records to better understand the connection between aerosol pollutants and health problems. In the context of a respiratory pandemic, quick, detailed, and accurate information about air quality can provide important data in the push to understand the disease and forecast severity of symptoms of the people infected by COVID-19. Another EO strength is the visual presentation of the data it collects. Striking images capturing air quality improvements during initial lockdowns have been widely disseminated (Figs 2.6.6 and 2.6.7) (Diffenbaugh *et al.* 2020).

Mobility is intimately connected with air quality, since vehicle emissions are a major source of air pollution, especially in urban areas. During the lockdown, night light imagery revealing darkened roads and satellite data recording NO_2 emissions have dramatically illustrated the link between car traffic density, air pollution, and ultimately human health. In fact, human-produced aerosol content in the air causes more light to be scattered, compounding light pollution at night; the reduced road and air traffic during lockdowns and the ensuing cleaner air contributed to decreasing artificial skyglow in many parts of the world (Bustamante-Calabria *et al.*, 2020; Jechow and Hölker, 2020). Some research has already come out quantifying the effects of air quality and disease severity (Wu *et al.*, 2020). One study compared air quality readings from 355 municipalities in the Netherlands, including data on nitrogen dioxide, sulfur dioxide, and fine PM. The team found that areas that had even slightly higher pollutant levels tended to have more cases, hospitalizations, and deaths (Cole *et al.*, 2020).

Air quality and the possible interactions between air pollution and the increasing risk of COVID-19 health impacts have garnered a lot of attention during the pandemic (Conticinia *et al.*, 2020; Travaglio *et al.*, 2020; Ogen, 2020). Because air quality is worse in cities, it is difficult to know if or to what degree the high rates and severity of infections are due to things like high density living during a pandemic that can be spread through the air or if it is also due to higher levels of aerosols in that air. Studies need to correct for spatial spillover (Cole *et al.*, 2020) and confounding factors like social deprivation, population density, ethnic composition, age, etc., but the connection is compelling. Conticinia *et al.* (2020) emphasize that prolonged exposure to air pollution leads to a chronic inflammatory stimulus and that northern Italy, with one of the highest death rates of the virus, also has some of the highest levels of air pollution in Europe (ESA, 2020) (Fig. 2.6.8).

Land use and land cover

The data EO satellites collected during the initial shutdowns can be used to track any changes in natural and protected areas, for example, to see if reduced human traffic had an effect on erosion, water quality, changes in wildlife behavior, or even invasive species. This can deepen our understanding of human impact on the planet and the actions that might be required to address issues like climate change. The pandemic did not just create an environmental reprieve, however. There has also been increasing animal poaching given food shortages and the collapse of tourism, which increases potential contact

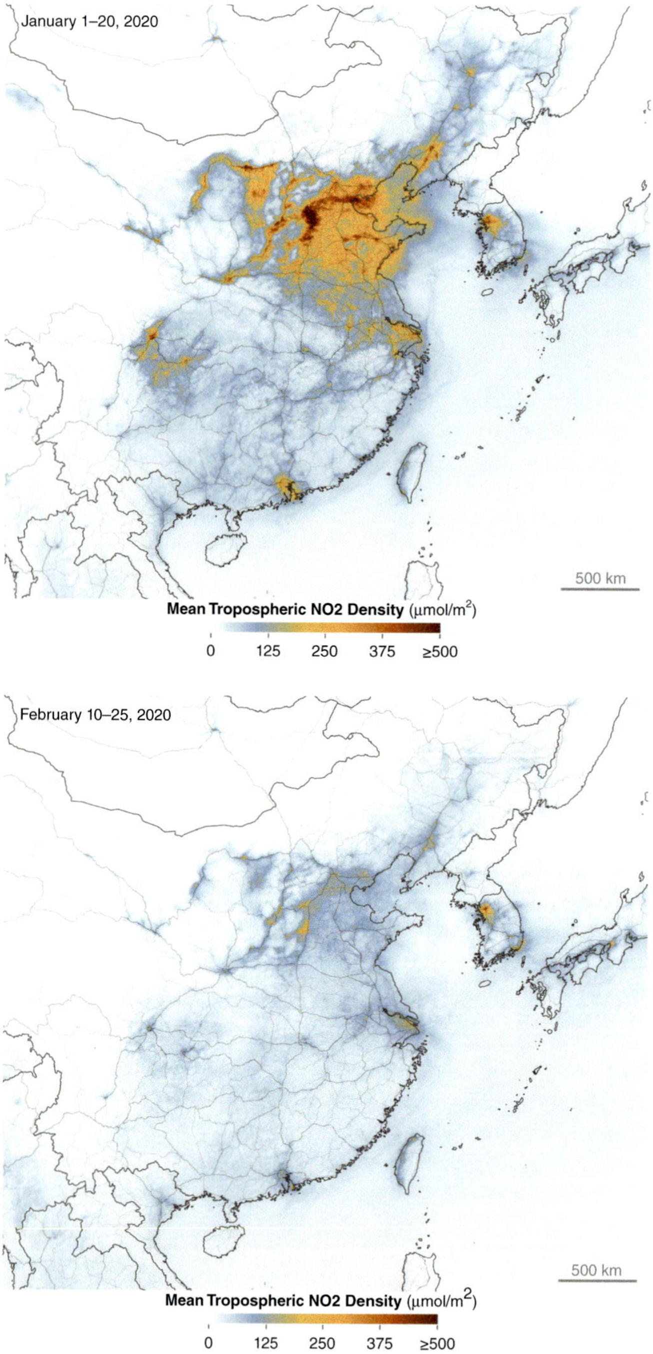

Fig. 2.6.6. Distribution of pre-COVID-19 lockdown nitrogen dioxide (NO_2) levels over eastern China 1–20 January 2020 based on data collected by the Tropospheric Monitoring Instrument (TROPOMI) on the European Commission's Copernicus Sentinel-5P satellite, and reduced NO_2 levels 10–25 February 2020 during the lockdown period, prior to lifting of quarantine restrictions and resumption of economic activities. (Image: ©NASA 2020.)

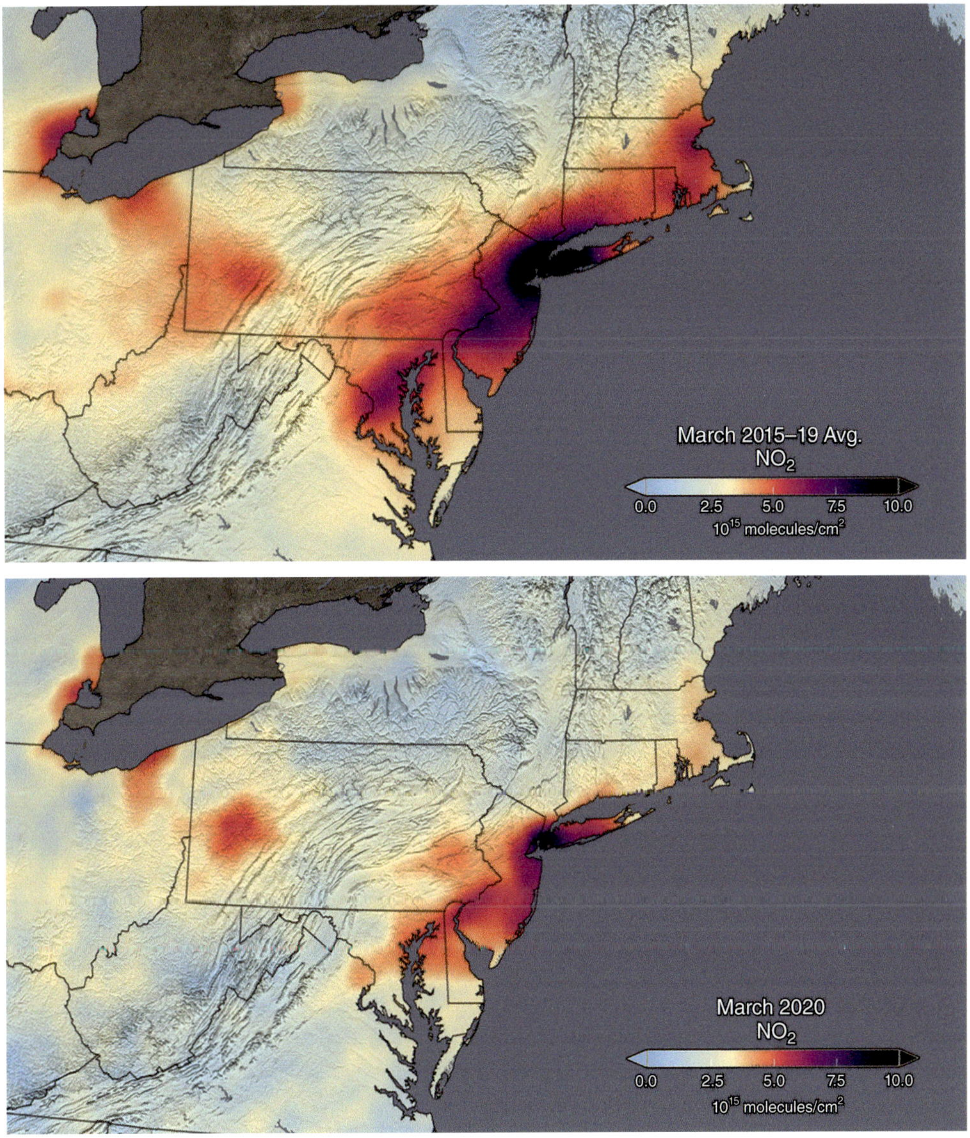

Fig. 2.6.7. NASA's Aura satellite measurements revealing significant reductions in nitrogen dioxide (NO_2) air pollution over the major metropolitan areas of the northeastern USA during the COVID-19 lockdown in March 2020 (bottom map) relative to the average concentrations during that month for the 2015–2019 period (top map). (Image: ©NASA 2020.)

between humans and zoonotic incubators, and a surge in agricultural expansion, logging, and illegal mining (United Nations Environment Programme and International Livestock Research Institute [UNEP/ILRI], 2020). According to Brazil's National Institute of Space Research (INPE), satellite data showed that 2020 saw more land cleared than 2019, which was already the biggest year for deforestation in more than a decade in Brazil.[11] More images are surfacing from agencies like ESA that show increased deforestation in the region. Because clearing the forest for agriculture or ranching is often done with fire and because the fragmentation of forested

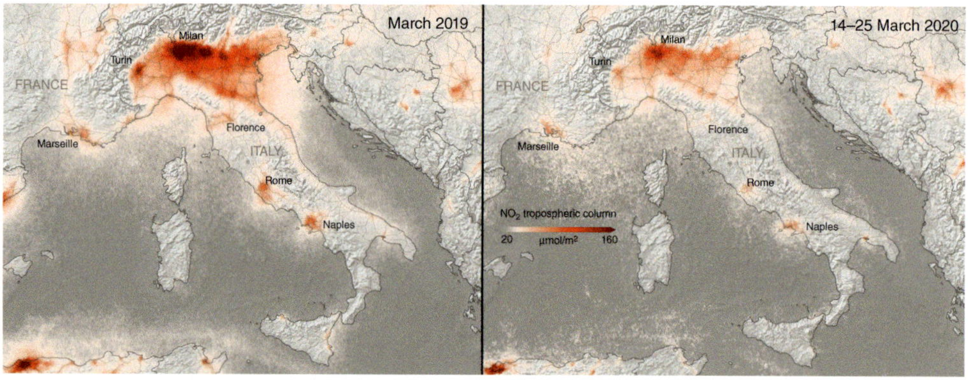

Fig. 2.6.8. Comparison maps of average nitrogen dioxide concentration over Italy as determined by European Copernicus Sentinel-5P satellite data analysis during the month of March 2019 (left) and during the period of 14–25 March 2020 (right) with marked reduction of concentration being attributed to reduced air pollution during COVID-19 lockdown measures. (Image: ©ESA, 2020. Contains modified Copernicus Sentinel data [2019–2020], processed by KNMI/ESA.)

land cover causes drier conditions (Gross, 2017), the fire season is predicted to be even worse in the future. This spells trouble for the severity of the pandemic in these areas. Wu *et al.* (2020) found that long-term average exposure to an increase of even 1 µg/m^3 of fine PM ($PM_{2.5}$) is associated with an 8% increase in the risk of dying from COVID-19. Deforestation and this pandemic are intimately connected and reducing forest loss is clearly a public health matter, now more than ever. Contributions of EO to issues of land cover and land use in South America include monitoring and assessing changes in forest cover, detecting negative changes in soil moisture for early warning of fire risk, and tracking the movement of smoke from wildfires for public health alerts.

Vulnerable populations

Given how accessible and reliable EO air pollution data are, it is tempting to see a large correlation between pollution and poor outcomes. However, a heterogeneity of risk across all individuals and community identities must be taken into account when predicting outcomes. Risk mapping is a common preparation and prevention activity. When an outbreak occurs, mapping risk takes on an immediate value and an urgency to inform response to an infectious disease outbreak. But without thinking about risk and exposure in a specific, targeted way, our risk models could miss the mark. Much has been written in 2020 about the heterogeneities of COVID-19 risk, as they include factors connected with age, sex, comorbidities, employment, Indigenous and ethnic identities, and structural barriers to health care (Mishra *et al.*, 2020; Schwalbe, 2020). India had one of the most stringent lockdowns but ballooning infection rates within weeks of reopening have to be accounted for by risk factors unique to the country or simply risk factors that are more prevalent or impactful. These may include willingness and ability to physically distance, population density, cultural beliefs, education, and of course the ability of the government and public health to respond to the crisis (Laxminarayan *et al.*, 2020). With this newly gained understanding of the heterogeneities of risk, EO could directly or indirectly address the following risk layers that impact vulnerable populations and the severity of outcomes: environmental (e.g. air quality), seasonal (weather, temperature), locational (e.g. exposure, access), mobility, and socio-economical (e.g. housing, density).

Looking ahead toward the implementation of COVID-19 related vaccination programs and measurements of their effectiveness, especially with regard to vulnerable and mobile populations, there is application potential for EO as well. Recent immunization-related studies prior to the pandemic have focused on low-income settings in Africa. Bharti *et al.* (2016) found that vaccination and immunization campaigns often did not reach their goal of obtaining accurate estimates of target populations and achieve coverage because of uncertainties concerning population size and distribution. Their research combined EO measurements of fluctuations in population distribution with high-resolution disease outbreak reports to guide potential improvement in vaccination campaign coverage plans. Models involving the use of satellite imagery demonstrated that retrospective estimates of vaccination campaign impact and future campaign planning can be improved by synchronizing interventions with predictable population fluxes. Employing EO and GIS for rural settlement status assessments as a guide for polio immunization programs in Nigeria, Higgins *et al.* (2019) presented strong evidence that this method has the potential to improve planning and implementation of public vaccination initiatives for at-risk populations trapped in conflict-affected regions around the world. The United Nations International Children's Emergency Fund (UNICEF) offered managerial considerations for in-country guidance on and strengthening of the use of geospatial data and technologies in immunization programs (UNICEF, 2018).

Conclusion

The scope and extent of the COVID-19 pandemic has challenged the capabilities of EO technology and the abilities of the EO community of practice in an extraordinary way. EO applications have been studied and adopted for a number of public health themes in the recent past. However, researchers and practitioners alike have conferred little attention to potential applications of EO in the case of a pandemic. The COVID-19 outbreak provided the opportunity to appraise actual EO responses during the first half of 2020. Although many of these responses are still tentative, the near-instantaneous appraisal from a public health perspective remains preliminary. While this review relies on both rapidly prepared, reviewed, and published papers and on authoritative online sources reporting on the utility of EO during the COVID-19 pandemic, there is a need for a comprehensive scoping review and in-depth analysis. This appraisal has yielded a number of preliminary conclusions, as follows.

The pandemic has galvanized EO players, satellite operators, and the GEO community of public health practice to provide geospatial information for the efforts to contain the virus and monitor socio-economic ramifications. The community was able to rely on regular EO satellite operations and data delivery services that functioned nominally and uninterrupted during the initial and widespread COVID-19 lockdown period.

Several routine EO applications have proven useful: assessing environmental changes; tracking compliance levels to public health measures and best practices; monitoring before, during, and after certain COVID-19 interventions and their effects on the environment and the economy; and providing safety- and security-related surveillance capabilities to public health-related GIS applications.

EO data have gained prominence in cartographic products and web-based dashboards detailing disease spread and its effects around the world. EO imagery has offered compelling visual information to an expanding audience of engaged citizens. It has provided key information on air quality and changes in land use and weather to support research into the virus and possibly to provide early warnings of increased risk of spread and disease severity for preparedness and early interventions.

EO data can contribute to assessing the risk of emerging health threats, modeling and mapping outbreaks, and supporting situational awareness and surveillance. Furthermore, EO can rapidly update data on land use, weather, night lighting, mobility, and air quality, thus lending geospatial–temporal dimensions to the heterogeneities of risk, health data, economic data, environmental data, and population distribution. Initiated by the ESA and the European Commission, the online dashboard of rapid EO action on COVID-19 represents a clear demonstration of satellite imaging capabilities to monitor the environmental and socio-economic impacts of the pandemic in spatial as well as temporal detail.

EO data can potentially support GIS applications during an epidemic or pandemic in LMICs where reliable, electronic health record systems, timely data for environmental monitoring, and up-to-date base map information may be intermittent or lacking. There is a general need to locate and support vulnerable populations; this may require data that GIS can quickly gather, analyse, and efficiently share. The effectiveness of COVID-19-related vaccination programs could potentially be improved by applying up-to-date EO and GIS data to assist in locating and immunizing vulnerable populations.

The cooperation of space agencies, EO data and service providers, and civilian authorities and the ad hoc EO applications regarding COVID-19 impacts have focused predominantly on environmental monitoring and public safety and security issues. In the absence of an EO pandemic playbook, the activities and developments to date represent useful demonstrations of EO capability rather than planned responses to specific geospatial requirements.

Satellite imagery can help civilian authorities plan for pandemic responses and recovery phases. Generating actionable knowledge and public information regarding the dynamics of the COVID-19 pandemic will be an urgent matter over the next few years. Research concerning key EO data collections and curatorial services will be essential for providing evidence-based data, information, and know-how for decision makers in the public health community of practice.

Acknowledgments

The contribution of Kaarla Sundström to the initial preparation of this appraisal of EO and the COVID-19 pandemic is gratefully acknowledged. Thank you also to Dr. Nick Ogden and Dr. Serge Olivier Kotchi for offering their valuable comments and suggestions.

Notes

[1] https://www.esa.int/About_Us/Business_with_ESA/Join_the_second_online_Global_Space_Economic_Workshop (accessed 4 January 2022).

[2] https://earthobservations.org/covid19.php (accessed 4 January 2022).

[3] https://eodashboard.org; https://race.esa.int/ (accessed 4 January 2022).

[4] https://www.spaceappschallenge.org/ (accessed 4 January 2022).

[5] https://experience.arcgis.com/experience/685d0ace521648f8a5beeeee1b9125cd (accessed 4 January 2022).

[6] https://app.developer.here.com/coronavirus/ (accessed 4 January 2022).

[7] https://storymaps.arcgis.com/stories/4fdc0d03d3a34aa485de1fb0d2650ee0 (accessed 4 January 2022).

[8] https://nymag.com/intelligencer/2020/04/jhus-coronavirus-site-gets-1-2-billion-interactions-a-day.html (accessed 4 January 2022).

[9] https://blog.maxar.com/news-events/2020/a-message-from-ceo-dan-jablonsky-on-maxars-response-to-covid-19 (accessed 4 January 2022).

[10] https://public.wmo.int/en/events/meetings/covid-19-symposium/outcomes (accessed 4 January 2022).

[11] http://www.obt.inpe.br/OBT/assuntos/programas/amazonia/prodes (accessed 4 January 2022).

References

Andersen, K.G., Rambaut, A., Lipkin, W.I., Holmes, E.C., and Garry, R.F. (2020) The proximal origin of SARS-CoV-2. *Nature Medicine* 26, 450–452. DOI: doi.org/10.1038/s41591-020-0820-9.

Bharti, N., Djibo, A., Tatem, A., Grenfell, B.T., and Ferrari, M.J. (2016) Measuring populations to improve vaccination coverage. *Scientific Reports* 6. DOI: https://doi.org/10.1038/srep34541.

Bustamante-Calabria, M., Sánchez de Miguela, A., Martín-Ruiz, S., J.L. Ortiz, J.L., Vílchez, J.M. *et al.* (2020). Effects of the COVID-19 lockdown on urban light emissions: Ground and satellite comparison. *arXiv: Instrumentation and Methods for Astrophysics*. DOI: https://doi.org/10.20944/preprints202011.0578.v1.

Cole, M.A., Ozgen, C., and Strobl, E. (2020) Air pollution exposure and COVID-19. IZA Institute of Labor Economics. Available at: http://ftp.iza.org/dp13367.pdf (accessed 3 August 2021).

Conticinia, E., Frediania, B., and Carob, D. (2020) Can atmospheric pollution be considered a co-factor in extremely high level of SARS-CoV-2 lethality in northern Italy? *Environmental Pollution* 261. DOI: https://doi.org/10.1016/j.envpol.2020.114465.

Diffenbaugh, N.S., Field, C.B., Appel, E.A., Azevedo, I.L., Baldocchi, D.D., *et al.* (2020) The COVID-19 lockdowns: A window into the Earth System. *Nature Reviews: Earth and Environment* 1, 470–481. DOI: https://doi.org/10.1038/s43017-020-0079-1.

Dong, E., Lu, H., and Gardner, L. (2020) An interactive web-based dashboard to track COVID-19 in real time. *The Lancet: Infectious Diseases* 20, 533–534. DOI: https://doi.org/10.1016/S1473-3099(20)30120-1.

Donnelly, C.A., Boyd, I., Campbell, P., Craig, C., Vallance, P. *et al.* (2018) Four principles to make evidence synthesis more useful for policy. *Nature* 558, 361–364. DOI: https://doi.org/10.1038/d41586-018-05414-4.

Elvidge, C.D., Hsu, F.-C., Ghosh, T., and Zhizhin, M. (2020) World tour of COVID-19 impacts on nighttime lights. Payne Institute for Public Policy. Available at: https://payneinstitute.mines.edu/wp-content/uploads/sites/149/2020/04/World-Tour-of-COVID-19-Impacts-on-Nighttime-Lights-3.pdf (accessed 3 August 2021).

Esri (2011) Geographic information systems and pandemic influenza planning and response. Available at: https://www.esri.com/library/whitepapers/pdfs/gis-and-pandemic-planning.pdf (accessed 16 June 2021).

Esri (2020) Geographic information systems for coronavirus planning and response. Available at: https://www.esri.com/content/dam/esrisites/en-us/media/pdf/geographic-information-systems-for-coronavirus-planning-response-white-paper.pdf (accessed 2 August 2021).

European Space Agency (ESA) (2020) Coronavirus: Nitrogen dioxide emissions drop over Italy. Available at: https://www.esa.int/ESA_Multimedia/Videos/2020/03/Coronavirus_nitrogen_dioxide_emissions_drop_over_Italy (accessed 16 June 2021).

Ferretti, L., Wymant, C., Kendall, M., Zhao, L., Nurtay, A. *et al.* (2020) Quantifying SARS-CoV-2 transmission suggests epidemic control with digital contact tracing. *Science* 368. DOI: https://doi.org/10.1126/science.abb6936.

Franch-Pardo, I., Napoletano, B.M., Rosete-Verges, F., and Lawal, B. (2020) Spatial analysis and GIS in the study of COVID-19. A review. *Science of the Total Environment* 739. DOI: https://doi.org/10.1016/j.scitotenv.2020.140033.

GHRF Commission, National Academy of Medicine (2016) *The Neglected Dimension of Global Security: A Framework to Counter Infectious Disease Crises*. The National Academies Press, Washington, DC. DOI: https://doi.org/10.17226/21891.

Gibney, E. (2020) Coronavirus lockdowns have changed the way Earth moves. *Nature*. Available at: https://www.nature.com/articles/d41586-020-00965-x (accessed 16 June 2021).

Gross, M. (2017) Brazil's fragmented forests. *Current Biology* 27. DOI: https://doi.org/10.1016/j.cub.2017.07.001.

Higgins, J., Adamu, U., Adewara, K., Aladeshawe, A., Aregay, A. *et al.* (2019) Finding inhabited settlements and tracking vaccination progress: The application of satellite imagery analysis to guide the immunization response to confirmation of previously-undetected, ongoing endemic wild poliovirus transmission in Borneo State, Nigeria. *International Journal of Health Geographics* 18. DOI: https://doi.org/10.1186/s12942-019-0175-y.

Jechow, A., and Hölker, F. (2020) Evidence that reduced air and road traffic decreased artificial night-time skyglow during COVID-19 lockdown in Berlin, Germany. *Remote Sensing* 20. DOI: https://doi.org/10.3390/rs12203412.

Jonas, O. (2013) Pandemic risk. World Bank. Available at: https://www.worldbank.org/content/dam/Worldbank/document/HDN/Health/WDR14_bp_Pandemic_Risk_Jonas.pdf (accessed 16 June 2021).

Kifer, D., Dario Bugada, D., Villar-Garcia, J., Gudelj, I., Menni, C. *et al.* (2021) Effects of environmental factors on severity and mortality of COVID-19. *Frontiers in Medicine* 7. DOI: https://doi.org/10.3389/fmed.2020.607786.

Kituyi, M. (2020) Why the global science community must come and stay together beyond the coronavirus pandemic. United Nations Conference on Trade and Prosperity. Available at: https://unctad.org/en/pages/newsdetails.aspx?OriginalVersionID=2357 (accessed 16 June 2021).

Lapolla, P. and Lee R. (2020) Privacy versus safety in contact-tracing apps for coronavirus disease 2019. *Digital Health* 6. DOI: https://doi.org/10.1177/2055207620941673.

Laxminarayan, R., Jameel, S., and Sarkar, S. (2020) India's battle against COVID-19: Progress and challenges. *American Journal of Tropical Medicine and Hygiene* 103, 1343–1347. DOI: https://doi.org/10.4269/ajtmh.20-0992.

Liu, Q., Sha, D., Liu, W., Houser, P., Zhang, L. *et al.* (2020) Spatiotemporal patterns of COVID-19 impact on human activities and environment in mainland China using nighttime light and air quality data. *Remote Sensing* 12, 1576. DOI: https://doi.org/10.3390/rs12101576.

Ma, Y., Zhao, Y., Liu, J., He, X., Wang, B. *et al.* (2020) Effects of temperature variation and humidity on the death of COVID-19 in Wuhan, China. *Science of the Total Environment* 724. DOI: https://doi.org/10.1016/j.scitotenv.2020.138226.

Ma, Y., Pei, S., Shaman, J. Dubrow, R., and Chen, K. (2021) Role of meteorological factors in the transmission of SARS-CoV-2 in the United States. *Nature Communications* 12. DOI: https://doi.org/10.1038/s41467-021-23866-7.

Minetto, R., Segundo, M.P., Rotich, G., and Sarkar, S. (2020) Measuring human and economic activity from satellite imagery to support city-scale decision-making during COVID-19 pandemic. *Institute of Electrical and Electronics Engineers* 7. DOI: https://doi.org/10.1109/TBDATA.2020.3032839.

Mishra, S., Kwong, J.C., Chan, A.K., and Baral, S.D. (2020) Understanding heterogeneity to inform the public health response to COVID-19 in Canada. *Canadian Medical Association Journal* 192. DOI: https://doi.org/10.1503/cmaj.201112.

National Academies of Sciences, Engineering, and Medicine (2020) Rapid expert consultation on SARS-CoV-2 survival in relation to temperature and humidity and potential for seasonality for the COVID-19 Pandemic (April 7, 2020). The National Academies Press, Washington, DC. DOI: https://doi.org/10.17226/25771.

Ogen, Y. (2020) Assessing nitrogen dioxide (NO_2) levels as a contributing factor to coronavirus (COVID-19) fatality. *Science of the Total Environment* 726. DOI: https://doi.org/10.1016/j.scitotenv.2020.138605.

Peckham, R. and Sinha, R. (2017) Satellites and the new war on infection: Tracking Ebola in West Africa. *Geoforum* 80, 24–38. DOI: https://doi.org/10.1016/j.geoforum.2017.01.001.

Schwalbe, N., Lehtimaki, S., and Gutiérrez, J.P. (2020) COVID-19: Rethinking risk. *The Lancet: Global Health* 8. DOI: https://doi.org/10.1016/S2214-109X(20)30276-X.

Timpka, T., Eriksson, H., Gursky, E.A., Strömgren, M., and Holm, E. (2011) Requirements and design of the PROSPER protocol for implementation of information infrastructures supporting pandemic response: A nominal group study. *PLoS ONE* 6. DOI: https://doi.org/10.1371/journal.pone.0017941.

Travaglio, M., Yu, Y., Popovic, R., Selley, L., Leal, N.S. *et al.* (2020) Links between air pollution and COVID-19 in England. *Environmental Pollution* 268. DOI: https://doi.org/10.1016/j.envpol.2020.115859.

United Nations Environment Programme and International Livestock Research Institute (UNEP/ILRI) (2020) Preventing the Next Pandemic: Zoonotic diseases and how to break the chain of transmission. Available at: https://www.unep.org/resources/report/preventing-future-zoonotic-disease-outbreaks-protecting-environment-animals-and (accessed 3 August 2021).

United Nations International Children's Emergency Fund (UNICEF) (2018) Guidance on the use of geospatial data and technologies in immunization programs (appendices). Available at: https://www.unicef.org/media/58181/file (accessed 16 June 2021).

US Department of Health and Human Services (2015) The National Public Health and Medical Situational Awareness Strategy Implementation Plan (2015–2018). ASPR: Assistant Secretary for Preparedness and Response. Available at: https://www.phe.gov/about/OPP/Documents/phm-sa-ip-sept2015.pdf (accessed 16 June 2021).

US Department of Homeland Security (2006) Pandemic influenza preparedness, response, and recovery. Guide for critical infrastructure and key resources. Available at: https://www.dhs.gov/sites/default/files/publications/cikrpandemicinfluenzaguide.pdf (accessed 16 June 2021).

World Health Organization (WHO) (2005) WHO global influenza preparedness plan: The role of WHO and recommendations for national measures before and during pandemics. Available at: https://apps.who.int/iris/handle/10665/68998 (accessed 3 August 2021).

World Health Organization (WHO) (2016) Report of the Review Committee on the Role of the IHR (2005) in the Ebola Outbreak and Response. Available at: https://apps.who.int/gb/ebwha/pdf_files/WHA69/A69_21-en.pdf (accessed 21 June 2021).

Wu, X., Nethery, R.C., Sabath, M.B., Braun, D., and Dominici, F. (2020) Air pollution and COVID-19 mortality in the United States: Strengths and limitations of an ecological regression analysis. *Science Advances* 6. DOI: http://doi.org/10.1126/sciadv.abd4049.

Zhang, T., Wu, Q., and Zhang, Z. (2020) Probable pangolin origin of SARS-CoV-2 associated with the COVID-19 outbreak. *Current Biology* 30, 1346–1351, DOI: https://doi.org/10.1016/j.cub.2020.03.022.

3 Needs, Challenges, and Opportunities: A Review by Experts

Stéphanie Brazeau[1], Cécile Vignolles[2], Ramesha S. Krishnamurthy[3], Juli Trtanj[4], John Haynes[5], Steven Ramage[6], Thibault Catry[7], Serge Olivier Kotchi[1], Marion Borderon[8], Michael Gill[9], Nicholas H. Ogden[1], Antoinette Ludwig[1], Guy Aube[10], Jan C. Semenza[11], Joaquin Trinanes[12], and Didier Davignon[13]

[1]Public Health Risk Sciences Division, National Microbiology Laboratory, Public Health Agency of Canada, Saint-Hyacinthe, Québec, Canada; [2]Centre National d'Etudes Spatiales, Directorate for Innovation, Applications and Science, Earth Observation Programme, Toulouse, France; [3]Health Systems and Innovation Cluster, World Health Organization, Geneva, Switzerland; [4]National Oceanic and Atmospheric Administration (NOAA) Climate Program Office, Bethesda, Maryland, USA; [5]NASA Earth Science Division, Washington, DC, USA; [6]Group on Earth Observations (GEO) Secretariat, Geneva, Switzerland; [7]ESPACE-DEV, Univ. Montpellier, IRD, Univ. Antilles, Univ. Guyane, Univ. Réunion, Montpellier, France; [8]Department of Geography and Regional Research, University of Vienna, Austria; [9]NatureServe, Arlington, Virginia, USA; [10]Space Utilization, Canadian Space Agency, St-Hubert, Canada; [11]Scientific Assessment, SRS, European Centre for Disease Prevention and Control (ECDC), Sweden; [12]University of Santiago de Compostela, Technological Research Institute, Santiago de Compostela, Spain/University of Miami, Cooperative Institute for Marine and Atmospheric Studies, Miami, Florida, USA/NOAA, Atlantic Oceanographic and Meteorological Laboratory, Miami, Florida, USA; [13]Canadian Meteorological Centre Operations, Environment and Climate Change Canada, Montréal, Canada

---

This section is an analysis of the needs and opportunities arising from international experts and managers in the field of Earth observation and publich health and the reference documents provided for this book. All the information collected has been grouped together into eight categories: (i) aligning with and supporting UN Sustainable Development Goals; (ii) focusing on public health needs and key theme areas for further research; (iii) accessing and developing Earth Observation (EO) and geospatial evidence-based data and products leveraging public health capacities; (iv) developing a sustainable community of practice; (v) developing knowledge and know-how; (vi) developing solutions: methods, tools, and systems; (vii) implementing technical infrastructures and technologies; and (viii) participating in EO satellite mission development for monitoring disease risks.

DOI: 10.1079/9781800621183.0003

Aligning with and Supporting UN Sustainable Development Goals

On the international stage, the United Nations Member States have adopted an action plan called *Agenda 2030 for Sustainable Development* that focuses on improving the lives of people and ending environmental degradation. The Agenda sets out 17 ambitious Sustainable Development Goals and 169 targets that integrate and promote transformative economic, social, and environmental activities of critical importance for humanity and the planet over the coming years.[2] Using remote sensing to aid public health aligns well with Sustainable Development Goal 3 (SDG 3), which aims to "ensure healthy lives and promote well-being for all at all ages." More specifically, EO and public health can offer promising support for Target 3D, which undertakes to "strengthen the capacity of all countries, in particular developing countries, for early warning, risk reduction and management of national and global health risks." However, remote sensing can support the implementation of a number of SDGs that are synergistic with SDG 3. For example, better biodiversity conservation efforts under SDG 15 will ensure sustainable provisioning services for humans, ecosystems, and, therefore, the mitigation of the risks of zoonotic diseases, both of which contribute to SDG 3.

The Secretariat of the international Group on Earth Observations (GEO) has developed an umbrella initiative to support the SDGs in a comprehensive way. The *EO4SDG* initiative identifies specific SDG targets for which EO can contribute development and progress indicators. The Secretariat is also trying to better understand the interactions between different SDGs, to develop a focus on health, and to increase policy opportunities for geospatial data.[3] Their EO toolkit for sustainable cities and human settlements (SDG 1)[4] is the first of multiple toolkits planned for policy and decision makers, executive managers, and the interested public to encourage awareness of relevant EO applications and to facilitate collaboration with EO experts to meet Sustainable Development Goals and contribute to One Health solutions to public health concerns. The World Health Organization (WHO) promotes the development of a supporting framework and seeks to optimize the combined use of EO data, routine health information data, and other remote sensing data for advancing target-specific SDG 3 activities at national and subnational levels. However, national ownership, intersectoral collaboration, and having the technical infrastructure, a competent workforce, and adequate finances are essential prerequisites for the framework to function effectively with EO and other geospatial data.

Focusing on Public Health Needs and Key Theme Areas for Further Research

The emergence of diseases poses a great challenge for public health. An important element in facing this challenge is the necessity of predicting and targeting the location and time when the risk of disease and the factors leading to it pose a threat to human populations and where and when it will spread. Research and development, surveillance activities, and operational capabilities must constantly adapt and evolve in response to this growing threat, particularly regarding obtaining information on changes in climate, in the environment, and in human populations[5] that drive disease emergence and pandemics. With greater knowledge of drivers of disease emergence, public health organizations can better predict, anticipate, and detect risks, thus allowing preparedness of disease prevention, prioritization of surveillance efforts, and control actions to mitigate the risks of disease exposure or transmission. Methods that provide information on disease risks are numerous. Anticipatory methods include risk assessment by predictive modeling and forecasting, while emerging diseases and pathogens are identified in laboratory or field surveillance, which may occur proactively or in the face of outbreaks. Field data, supported by accurate laboratory diagnosis, are critical for the calibration and validation of models. This entails the collection of data on disease cases reported, socio-economic and demographic data, as well as environmental and climate data. In this context, EO data analysis becomes particularly useful as it supports the understanding and integration of disease emergence drivers as part of the modeling process. For example, by improving our understanding of wild animal vectors (such as bats) and their relationship with habitat features, we can use EO to monitor and model various climate and development scenarios to predict future areas vulnerable to zoonotic disease outbreaks.

Several research themes were identified as important for public health and where EO has the highest potential of having its greatest effect. They include mosquito-borne diseases, tick-borne diseases, air-borne diseases (pollution and extreme heat), water-borne diseases, vulnerable human populations, and pandemics and major outbreaks such as COVID-19. In North America and in Europe there are several organizations and programs that support R&D and offer specific theme-related solutions. Examples include the following:

- In Canada, the National Microbiology Laboratory, Public Health Risk Science Division of Public Health Agency of Canada (PHAC) provides EO solutions, R&D products from spatial modeling, surveillance products, services for the Canadian public health community, and support to public health emergency centers.
- The Canadian Space Agency's Space Utilization Grant and Contribution programs support the Canadian government priorities by funding industry and academia for many thematic application areas, including public health.
- In the USA, NOAA's One Health program provides EO solutions to identify heat and health threats, air pollution, and other related topics.
- NOAA's International Research and Applications Project, IRAP, supports activities to link science and assessments to practical risk management challenges in regions where weather and climate affect US interests at home and abroad; IRAP priorities in 2018 included "Decision Support Research on Climate-Sensitive Health Risks."
- NASA's Health and Air Quality program supports the use of EO in air quality management and public health, with emphasis on infectious diseases and environmental health issues.
- In France, Centre National d'Études Spatiales' (CNES) programs support research laboratories in the field of tele-epidemiology.
- EcoHealth Alliance, based in New York City, researches, monitors, and predicts the emergence of new zoonotic diseases through *in situ* data collection and modeling of the relationships between habitat conversion, wildlife trade, human population patterns, and other key factors to predict and better understand the factors driving the emergence of pandemics.

Accessing and Developing EO and Geospatial Evidence-based Data and Products Leveraging Public Health Capacities

Access to timely and accurate geospatial data in support of effective evidence-based decision making is one of the challenges facing public health organizations. The emergence of diseases, which is influenced by drivers such as changes in human or animal behavior, the environment, or the climate, has significant geospatial dimensions (Fig. 3.1). Geospatial information is needed for the development of indicator and risk models related to: human ecology, including the distribution, abundance, vulnerability, and behavior of humans; land use, land cover, and land degradation; human, animal, and arthropod vector habitats and biodiversity; the role of certain animal species that can amplify the threat and spread of disease; and mechanisms that interconnect all of the above. Geographic aspects are also important in that certain drivers have different impacts on and significance for disease dynamics in different parts of the world. Location-specific data, then, are vital in addressing risk, and EO is a meaningful source for these data for all regions of the globe. Satellites provide data that are continuously archived and cover most of the Earth, including remote and hard-to-access regions. Using multitemporal EO data makes it possible to update the information as needed. Depending on the satellite and sensor system selected, daily, weekly, monthly, or seasonal EO-based updates can be provided.

In order to meet their EO data requirements and utilize the data effectively, public health organizations need to collaborate with space agencies and other organizations that provide access to EO missions and data streams. Several public and private organizations offer procedures for ordering new satellite imagery and retrieving archival data sets. Selection and access to data

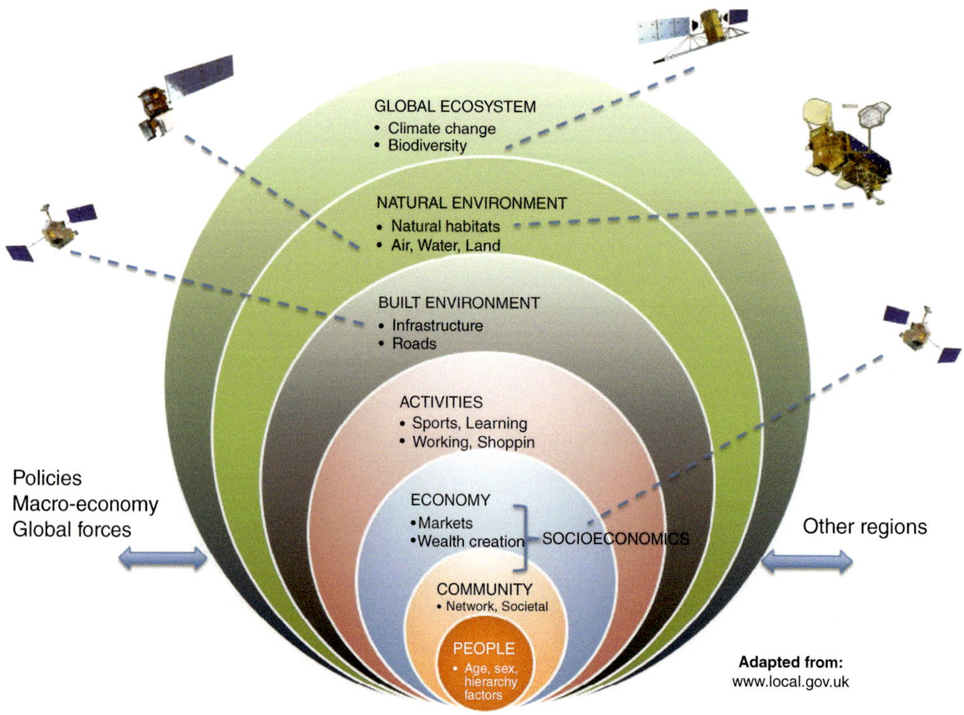

Fig. 3.1. EO and the multiple dimensions of health determinants.

and capacity to develop thematic information generated from those data can meet a substantial number of geospatial information needs for public health research activities. Near-real-time satellite data streams for generating up-to-date geospatial information are critical in addressing health-related emergency/emerging situations. Examples of high-resolution imagery for local and regional use include the Canadian RADARSAT constellation and the European series of Sentinel satellites under the Copernicus Programme; likewise, near real-time moderate-resolution data offered by US missions could satisfy information needs at the national or international scale. Additional EO data products (e.g. Global Forest Change data; Hansen *et al.*, 2013) can also prove useful for modeling disease emergence risk based on areas undergoing rapid land use change.

Optical and synthetic aperture radar (SAR) sensors at very high resolution are very useful for studies such as those investigating mosquito-borne diseases or COVID-19 disease within urban environments. In a research context, these types of studies are feasible but in an operational context, the cost and volume of such data could be prohibitive. Open access to EO data is critical. Sources for satellite-generated climate or environmental data that can be used to help assess the exposure of populations to a disease with environment and/or climate drivers are varied, some of which are freely available online.[6] A list of EO sources is included in Quattrochi *et al.* (2017) and in-operation open access EO images relevant for health determinants as well as advantages and limitations of the use of these images for health studies are included in Kotchi *et al.* (2019). Open data and data-sharing policies and promotion of participatory approaches to generating and accessing geospatial information are important prerequisites, as is the collection of health-related *in situ* data for producing spatial analyses.

Health studies and risk assessments rely on a variety of data sources (e.g. demographic, socio-economic, environmental) that must be integrated into models and health systems; providing reliable and consistent results remains essential. When dealing with health problems in different locations around the world, data sources can be heterogeneous in terms of content and quality. However, EO data obtained and used with appropriate analysis methods produce effective, homogeneous, and standardized information. Expanding and improving the accuracy of EO data will enhance model precision, sensitivity, and capacity to adapt to small changes in drivers that could influence the emergence of outbreaks.

The type of health issues and information needed will dictate public health decisions regarding the most appropriate resolution of EO images to use. Most mosquito-borne diseases require a high spatial resolution of ≤30 m to identify variations in the environment and climate used to support public health decision making. In addition, the availability of EO images makes it possible to develop risk and vulnerability maps at fine spatial and temporal resolutions. Water-borne disease risk modeling such as for *Vibrio* infections depends on EO images that support the modeling of sea surface temperature and sea surface salinity data. These data are available at low spatial resolution. The presence of *Vibrio* species in the water requires daily monitoring over time. EO data of greater spatial resolution could help refine these types of models. Air quality and heat wave modeling also uses the low-spatial-resolution data that are currently available, but high spatial resolution with frequent revisits is needed to better characterize urban pollution. COVID-19 has taught us how important data on inhabited environments are to assessing the effectiveness of public health measures and policies such as lockdowns or the links between the severity of symptoms and air quality. The availability of fine-resolution and timely images will also make it possible to make associations between the mobility of people and places at risk.

Local-scale risk prediction and surveillance are needed for differentiating objects or environmental changes in highly heterogeneous environments and for supporting local-level management. The timeliness of EO data is also an important factor because using these data in rapid risk assessment models or forecast models requires them to be both as recent as possible and ready to use. Downscaling approaches could provide the opportunity to obtain data with both high spatial and temporal resolution. They could involve multi-sensor data sets that integrate images of low spatial resolution and very high temporal resolution (daily) with multi-spectral images of high spatial resolution. Effective observation by air-borne and satellite-based optical sensors may at times be affected by clouds, pollution, and smoke, thus generating missing data and reducing the usefulness of EO products. Combining SAR and optical data offers some solutions to missing EO data due to atmospheric conditions.

Obtaining the desired data at all times and for all circumstances is often not possible. Although EO technologies are evolving rapidly, data useful for assessing a health event may be missing or may be obtained at unhelpful spatial or temporal resolutions. In coastal areas and in estuaries, the complexity of coastal waters makes it difficult to monitor water quality by EO satellite. This is the case for sea surface salinity data – currently available at a spatial resolution of 25 km – which affects coastal modeling of non-cholera *Vibrio* risk. Consequently, field instruments and modeling must make up for this lack of data. A multi-spectral or hyper-spectral sensor having the capacity to characterize surface salinity and surface temperature on a fine scale in coastal environments would be ideal.

In situ and epidemiological data

When studying a disease, combining geospatial data with other types of data such as *in situ* ecological, climate and weather, entomological, epidemiological, human demographic, socio-economic, and behavioral data is often necessary to calibrate models and validate their predictions. However, access to *in situ* data might be difficult because they are not always available, and the cost and time associated with collecting them is sometimes prohibitive.

In general, the more the better when it comes to accurate data on human cases and on presence, abundance, and persistence of microbes, arthropod vectors, relevant animal host populations, and human population characteristics. In low-income countries, researchers and public health practitioners depend on limited resources and often inadequate data on prevalence and incidence of the disease, including poor representation of the actual affected population (Programme National de Lutte contre le Paludisme, 2008; Diallo *et al.*, 2012), as well as other ground-level relevant geospatial data (Section 2.5).

Interdisciplinary and inter-sectoral collaboration is needed to identify and prepare relevant EO products for public health applications. Once such products and their specifications are identified in collaboration with experts in land use, land cover, or infrastructure assessments, funding agencies have the opportunity to support the development of products related to the key public health themes outlined in this section and to foster collaboration among organizations. Now and in the future, EO technologies hold promise for detecting and capturing attributes of the Earth at increasing detail, which will support the development of tools and models that will provide better information on health threats.

Developing a Sustainable Community of Practice

The development of a strong and vibrant community of practice concerned with the use of EO for public health purposes is at the foundation for innovative action. This may take place domestically and internationally and result in improvements in how geospatial information is obtained, adopted, and managed for public health issues. Networking and collaboration also reinforce communication about health risks and vulnerability of populations, scientific discoveries, and emerging technologies. The participants in the One Earth – One Health workshop, held in Montreal in 2017, proposed several approaches to building and sustaining an active community of practice:

- Identify and articulate the needs of public health stakeholders and end users.
- Identify country-specific health-related activities that are addressed or could be addressed within the geospatial domain in the following dimensions: health technical areas, resource needs, and research agenda.
- Encourage interdisciplinary and inter-sectoral cooperation of public health organizations, space agencies, academic institutions, and industry under the One Health approach.
- Encourage international joint activities with GEO and WHO, encourage national space agency and public health institution initiatives and framework development, and address common public health needs.
- Support, design, and implement public health-related outreach activities involving EO and other geospatial data.
- Identify and implement national capacity-building opportunities that can strengthen spatial and temporal coverage of *in situ* data (e.g. in collaboration with Group on Earth Observations Biodiversity Observation Network [GEO BON]).
- Support and participate in the Convention on Biological Diversity (CBD), particularly with regard to implementation of the post-2020 Global Biodiversity Framework.
- Encourage formal cooperative agreements and activities among health and EO communities within government organizations, academia, and industry at the international and national level.
- Create opportunities for regular dialogue between EO experts and researchers, managers, and practitioners in the field, and encourage open access knowledge sharing; for example, the GEO Knowledge Hub is a cloud-based digital library (repository) that shows all the components of a given EO application required to reuse it.[7]
- Form scientific teams to study and report on specific topics, such as the formal Health and Air Quality Applied Science Team (HAQAST) supported by NASA.
- Support and participate in the volunteer effort of the GEO Health Community of Practice (CoP).[8]

Efficient interaction between the remote sensing, entomology, ecology/biology, epidemiology, animal health, environmental science, climatology/meteorology, social sciences, and public health communities is essential to informing a CoP that can integrate data from a wide variety of sources and at various scales and qualities. The provision of data and expertise by the large remote sensing community has the potential to provide data at a range of spatial and temporal resolutions that can support public health activities. EO products and methodologies will initially have to be custom designed to better fit public health needs, and proven methodologies need to be automated and more robust in the future – as well as user-friendly enough to be implemented by non-specialists.

While the technology can be further improved and more extensively employed, geospatial data producers within the Canadian federal community are well networked and already producing relevant information to inform public health mandates. For example, their *Ixodes scapularis* risk map (Section 2.2) provides reference data of the quality and relevance needed to feed predictive models of vector-borne disease and risk maps. Given globalization and the interconnectedness of the planet, expanding this CoP will benefit public health initiatives both at home and around the world.

Developing Knowledge and Know-how

The multi-disciplinary approach promoted by the One Health concept will help increase the knowledge and know-how of stakeholders to better identify and address the information requirements for critical health issues. Chief among them is a better understanding of how climate, environment, biodiversity, and socio-economics – including human–animal–environment interactions – affect the potential occurrence and emergence of diseases at various scale levels. There is also the need to improve prediction of diseases, detect emerging hotspots, and build early warning systems.

EO technologies and geomatics do, or could, satisfy many of these information requirements, and building the capacity for skilled human resources is essential for effectively utilizing EO data and deriving geospatial information from them. The following list includes proposed methods to stimulate the development of knowledge and know-how in the inter-related fields of EO and public health:

- Use government-assisted programs to develop know-how for the acquisition of EO data, the analysis of EO-derived information, and the integration with other geospatial data to support priority areas of research.
- Strengthen arrangements between public health laboratories and academic institutions focused on remote sensing, geomatics, geography, and other public health interests to develop highly qualified personnel and to support future research.
- Strengthen collaboration among space agencies and EO organizations and national and international public health agencies to widely share EO knowledge and best practices.
- Support the professional development and academic programs for a new generation of EO and health specialists (MSc, PhD) and R&D activities related to emerging fields such as big data analytics and deep and machine learning.
- Offer training through applied remote sensing training programs (e.g., NASA applied remote sensing training program [ARSET]), workshops that offer training on relevant EO and geomatics skills for end users, and long term educational partnerships with communities and institutions in the public and private sectors.

Expertise in EO image analysis, geo-informatics, and mapping are prerequisites for the development of risk information products such as maps. There is a significant demand for assistance in the process of skills, knowledge, and technology transfer. This applies especially to the public health sectors of many countries that seek to employ effective geospatial assets to counter the threat of infectious diseases with the help of EO data analysis. Interagency cooperation to create new products that offer significant benefits to societal health is important. Health system improvements require dedicated resources for model development, EO data processing, and model forecast operations. In addition, skills and capacity among end users of integrated health and geospatial products are required to maximize their use.

Developing Solutions: Methods, Tools, and Systems

Public health organizations have been using the best available evidence and tools to advise and support national and international stakeholders in their work to enhance the health of their respective communities. However, more innovative scientific tools and methods need to be researched and promising solutions need to be implemented in public health programs to help combat increasing threats from infectious and chronic diseases. Hence, appropriate and interoperable EO-based products have to be specified, and spatial, analytical, and timely solutions need to be developed with the public health community. These could focus especially on epidemiological analyses, risk modeling, surveillance and investigation, and emergency management.

The EO-based information could support and improve public health decision making at many time scales, such as early warning forecasts for disease management of the most vulnerable areas and engaging in preparatory communications and planning for health administrators. Using EO data and tools, it could be possible to shift the current focus on responding to outbreaks toward predicting and preventing diseases.

Integrated health information systems offer solutions to address the current information gap between early warning and early action. The development of integrated methodological approaches from different fields of expertise using a wide range of relevant data to obtain public health risk maps serve to illustrate the complexity of public health issues. The French Space Agency (CNES) and its partners have developed a concept based on a deterministic/statistical approach of the climate–environment–health relationships adapted to what the space sector can offer; the approach is multi-disciplinary in that the study of the key mechanisms favoring emergence and propagation of infectious diseases brings together disciplines like environmental studies, climate science, social sciences, microbiology, entomology, and veterinary sciences.

In a similar way, GEO BON is fostering the development of an interoperable biodiversity observation system at national, regional, and global scales, and across terrestrial, marine, and aquatic systems. It integrates *in situ* and remotely sensed monitoring systems that bring together biodiversity, ecosystem conditions, and wildlife-related health observations. This system, in part, provides information on the change in biodiversity that could facilitate the emergence of infectious diseases and the exposure of vulnerable populations. GEO BON is working with the Open Geospatial Consortium[9] on interoperability across analysis-ready data (ARD) tools and services.

It is important to evaluate current systems that are intended to provide information about health risks in order to propose innovative ways to represent the level and spatial distribution of health risks. In this context, GEO's analysis of the resilience of the systems in place to inform health stakeholders would enable health and EO specialists to assess the capacity of the systems to perform this task. This evaluation of the systems would pinpoint both the weaknesses and the opportunities of the systems, both of which need to be considered when looking at how shocks and stresses affect systems and people. Depending on the solution, collaboration would help integration of health and environment data, different metrics, and reporting systems. While there is obviously a cost to this endeavor, the cloud offers a lot of opportunity around data infrastructure. In fact, GEO has 55 projects currently running with Amazon Web Services (AWS), Google Earth Engine, and Microsoft artificial intelligence (MS AI) for Earth.

Important EO systems include those provided by Copernicus,[10] Global Earth Observation System of Systems (GEOSS),[11] and Committee on Earth Observation Satellites (CEOS).[12] These systems could be assessed to measure how they could provide information on health risks and how they could support the integration of health systems. Health systems that could benefit from EO systems will have to be identified.

Artificial intelligence (AI) and Analyse ready data (ARD) solutions for complex issues

In an ever more complex world, where many different factors intersect to play dynamic and complex roles that affect the health and well-being of people, the ability to observe, measure, understand, assess, and take action on these determinants of health is becoming exponentially more complex. The methodologies for the production of risk maps developed by researchers of the EO community are not always suitable or adequate in a public health context for reasons that include the complexity of the methodologies, the cost of high-resolution data, and the lack of computing resources. Multi-temporal series of optical

EO data and the combination of optical and SAR data necessitate large data storage and analytics resources for regular production of risk maps. Also, the addition of new sensors (e.g. RADARSAT constellation, Surface Water Ocean Topography [SWOT], Biomass) increases the volume of EO data for their utilization while also compounding a data storage challenge. The development of adapted computing methods such as Artificial intelligence (AI) and machine learning algorithms with storage capacities will provide solutions that will need to be customized for public health purposes. The application of AI and related big data technologies could play a critical role in the enhanced application of EO to all the health-related activities discussed in this book (i.e. COVID-19). The potential benefits of applying AI and big data technologies to issues that influence health, safety, and well-being are, at this time, focused on critical zones where health priorities and this domain of innovative science and technology intersect.

The health community has identified the priority of increasing the capacity of public health officers to conduct rapid public health mapping and spatial analysis. To reach this goal, EO data management, a data cube, and a system that can provide Analysis ready data (ARD) for rapid modeling and timely risk mapping must be further developed. For example, Digital Earth Africa[13] offers continental water observations from space for free on an almost daily basis. Automated and generic methods are preferable to facilitate the production of EO-based products like land cover maps anywhere in the world. A high-performance computing system and cloud capacity will have to be studied to identify the best solution for big data storage and analytics and the best approach to produce health-related results. A large volume of data could be remotely processed and analyzed following the model proposed by Google Earth Engine – that is, without downloading data. Health communities would benefit from health systems that can process data with a secure interface, allowing the development of a sensitive and protected product. Public health and EO communities need to support research that would provide them with AI tools for big data analysis tailored to their needs.

Most EO data analysis and the production of risk maps require image processing time in expert software that can be done, in part, with geographic information system (GIS) software. Availability of freeware with EO and GIS tools, open access to EO data, as well as training programs strongly encourage the use of EO products. Future development should consider the implementation of tools through open-source software available internationally.

Implementing Technical Infrastructures and Technologies

The technical infrastructure for using EO data is a precondition of undertaking massive geospatial analyses to support public health-related decision making. Countries without this infrastructure depend on the infrastructure of other organizations to obtain EO data sets and to support their massive analyses. In some countries, the infrastructure to support big data analytics simply does not exist, or no formal agreement is in place to use existing infrastructure for EO and public health matters.

For those countries that currently have relevant infrastructure, the large volume of EO data streams and the high rate at which these require updating for disease risk assessment could rapidly exceed existing information management and information technology capacities and technologies. Producing and archiving data, products, and maps with high spatial resolution (≤30 m) for diseases that need this level of precision would require an exceptional data storage capacity. As the need for high-volume health-related data on environmental, climatic, and socio-economic factors has increased both domestically and internationally, the public health community is continuously challenged to maintain access to timely, reliable, and accurate EO data. Many health-related systems that integrate geospatial data already need big data infrastructure for storage and processing and sometimes 24/7 support for its operation, as is the case for AIRNow, FireWork systems, and the *Vibrio* map viewer (Chapter 2).

Updated and expanded IT infrastructure and software is a partial answer to the problems facing people responding to public health priorities and crisis. More effective and accurate mapping capacities support a variety of activities: risk assessments and decision making during health emergency events; risk communication via supplied images; and evaluation of factors affecting health via risk modeling. There is also a need for enhanced on-site information and the development of databases for the surveillance of diseases, so that greater effort can go toward the development of efficient EO-based models and tools to inform decision makers.

Participating in EO Satellite Mission Development for Monitoring Disease Risks

There are currently hundreds of EO satellites orbiting in space, and new missions are continually being planned for deployment. The usefulness of EO satellites depends in large part on the ability of users to access and apply the data and technology in practical settings to address their pressing issues. Satellite sensors are not primarily designed for health applications and often render spatial, temporal, or spectral data properties that are not of use for addressing public health issues. In Canada, the Canadian Space Agency has been building teams of remote sensing experts from different sectors, including public health. These users and science teams are integral parts of mission planning and utilization cycles, helping to identify observation requirements, specify technical needs, and develop instrument designs to meet a wide range of requirements and EO data needs. These teams can also participate in the calibration and validation of satellite data to ensure data quality.

Some EO satellite systems offer ARD (i.e. pre-processed images) and related information products derived from the raw data stream generated by the satellite instruments and the use of algorithms. For example, the Moderate Resolution Imaging Spectroradiometer (MODIS) sensors onboard the US Aqua and Terra satellites offer atmosphere, land, cryosphere, and ocean products that are used in several user communities.[14] MODIS indicator data sets have been so successful that they do not require additional remote sensing analysis; they can be used directly in predictive models. Some missions have been influenced to produce data and algorithms specific to monitoring and managing health issues; for example, the NASA Earth Venture Instrument-3, which is a new Multi-Angle Imager for Aerosols (MAIA[15]). MAIA's mission objective with regard to health is to assess links between different air-borne particulate matter types and adverse birth outcomes, cardiovascular and respiratory disease, and premature deaths. Participation in mission development is an excellent networking and collaboration opportunity that will lead to important advancements in the field. One such advancement attributable to Landsat data is the ability to monitor changing patterns in forest cover loss and human encroachment on previously wild areas that allows for better prediction of zoonotic disease emergence.

Notes

1 https://crss-sct.ca/conferences/csrs2017/one-earth-one-health-workshop/ (accessed 6 January 2022).

2 https://sustainabledevelopment.un.org/sdgs; https://sustainabledevelopment.un.org/post2015/transformingourworld (accessed 6 January 2022).

3 http://eo4sdg.org/ (accessed 6 January 2022).

4 https://eo-toolkit-guo-un-habitat.opendata.arcgis.com/ (accessed 6 January 2022).

5 https://ghsl.jrc.ec.europa.eu/HPI.php (accessed 6 January 2022).

6 https://earthobservations.org/open_eo_data.php (accessed 6 January 2022).

7 https://earthobservations.org/gkh_webinars.php (accessed 6 January 2022).

8 http://www.geohealthcop.org/ (accessed 6 January 2022).

9 https://www.ogc.org/ (accessed 6 January 2022).

10 https://www.copernicus.eu/en/about-copernicus (accessed 6 January 2022).

11 https://www.earthobservations.org/geoss.php (accessed 6 January 2022).

12 http://ceos.org/about-ceos/overview/; http://ceos.org/data-tools/; http://ceos.org/ard/ (accessed 6 January 2022).

13 https://www.digitalearthafrica.org/why-digital-earth-africa/water-resources-and-flood-risks (accessed 6 January 2022).

14 https://modis.gsfc.nasa.gov/data/ (accessed 6 January 2022).

15 https://maia.jpl.nasa.gov/ (accessed 6 January 2022).

References

Diallo, A., Ndam, N.T., Moussiliou, A., Dos Santos, S., Ndonky, A. *et al.* (2012) Asymptomatic carriage of *Plasmodium* in urban Dakar: The risk of malaria should not be underestimated. *PLoS ONE* 7. DOI: https://doi.org/10.1016/j.gloenvcha.2010.07.003.

Hansen M.C., Potapov P.V., Moore R., Hancher M., Turubanova S.A. *et al.* (2013) High-resolution global maps of 21st-century forest cover change. *Science* 342, 850–853. DOI: https://doi.org/10.1126/science.1244693.

Kotchi, S.O., Bouchard, C., Ludwig, A., Rees, E.E., and Brazeau, S. (2019) Using Earth observation images to inform risk assessment and mapping of climate change-related infectious diseases. *Canada Communicable Disease Report* 45, 133–142. DOI: https://doi.org/10.14745/ccdr.v45i05a04.

Programme National de Lutte contre le Paludisme (PNLP) (2008) Morbidite et Mortalite Palustre au Sénégal en 2008: Synthèse donneés districts & hospitaux. PNLP, Dakar, Sénégal.

Quattrochi, D.A., Wentz, E., Lam, N.S.-N., and Emerson, C.W. (eds) (2017) *Integrating Scale in Remote Sensing and GIS.* Routledge, New York.

4 Conclusions and Opportunities

This book compiles extensive information on the potential for Earth Observation (EO) to contribute to public health practice. Contributors include scientists, policy makers, and managers from both the EO and public health communities. This subsequent compilation of expertise along with contributions from the international community, is primarily intended to help managers interested in developing geospatial applications in the area of research, risk assessment, and early warning systems for public health. We hope that both communities find the book useful in deepening their commitment to working closely together.

Infectious and chronic diseases, whether on a global, regional, or local level, are public health concerns that affect individuals but pose escalating threats to the larger population. Because generating evidence-based knowledge about these diseases is key to managing them and reducing their impact, the capacity of EO satellites to quickly and accurately collect extensive data sets on the changing drivers for disease occurrence and spread gives public health a tactical advantage in predicting disease risks. To continue improving this capacity and to expand public health applications of EO technology to meet the information needs for a broader range of health risks – including its potential to help combat endemic, epidemic, and pandemic infectious diseases – future satellite-based EO derived products need to prioritize emerging public health needs. This would entail ensuring that public health information needs regarding climate, environmental, and human population changes are met. These data would also have multiple applications, with the ultimate goal of informing sustainable development and building human resiliency.

As the interactions between humans, farm animals, and fauna intensify – due to increased trade in bushmeat and wild species, human encroachment on wild habitats, changes in land use practices, and climate, ecosystem, and socio-economic changes – the risk of emerging infectious diseases also increases. The recognition of this has led to the widely adopted One Health approach, which promotes inter-sectoral collaboration to integrate human, animal, and environmental health. Interdisciplinary work and inter-sectoral collaboration are promising ways to facilitate data identification and encourage preparation of relevant EO products that can effectively meet public health challenges. EO experts working together with experts from a variety of sectors can push innovation on how and what data are collected and how they are processed and applied to these complicated challenges.

DOI: 10.1079/9781800621183.0004

Remote sensing can afford us a unique perspective on the Earth, helping us build our scientific understanding of the planet as a system. When combined with other sources of data like health, socio-economic, and demographic data, EO can be an effective tool in understanding, modeling, and forecasting public health threats.

In its 2030 Agenda for Sustainable Development, the United Nations (UN) identifies the Sustainable Development Goal (SDG) of ensuring good health and well-being for people of all ages by strengthening "the capacity of all countries, in particular developing countries, for early warning, risk reduction and management of national and global health risks." Remote sensing can play a well-defined role in working to meet the target objectives of the UN's SDG initiative. Major organizations like the World Health Organization (WHO) and the Group on Earth Observations (GEO) also point to the potential of collaboration between EO sectors and public health experts. With this in mind, this book addressed three basic questions: How does, or can, the current capacities of EO assist public health activities? What are the challenges for operational use of EO in public health? What opportunities are there to further develop EO to the future benefit of public health?

Priority themes for which EO appears promising have been identified in this book:

1. For infectious diseases: major epidemics and pandemics such as COVID-19, vector-borne diseases transmitted by mosquitoes and ticks, and water-borne diseases such as pathogenic *Vibrio* species.
2. For chronic illnesses: impact of air pollutants and extreme heat.
3. For all diseases: indicators of human populations at risk could be developed to target the most vulnerable and areas at greatest risk for new disease emergence.

The examples for each of these priority themes in this book show the usefulness and benefit of EO data as a strategic tool for assessing and monitoring public health risks in an effective and continuous way. EO images have been able to help identify disease risk areas in endemic countries or in newly emerging areas because they are able to detect land use practices, land cover, climate information such as land or sea surface temperature, and qualities of human urban, suburban, and rural environments. These examples demonstrate that EO data play an important role in developing risk models, which in turn are used to create program-ready risk communication tools such as risk maps. Models and maps make it possible to generate information on the occurrence, importance, and future likely distribution and spatial spread trends of risk, as well as on risk factors that may explain the presence or emergence of a disease. Ultimately, data and products derived from EO make it possible to identify and locate the risk factors in a given territory that determine if a disease or a vector is present or if there is a current risk, given suitable environmental and climate conditions.

Risk maps from EO images allow public health actors to anticipate and prepare for health threats because they can act as an early signal. Perhaps most significantly, because EO and risk maps can detect favorable conditions for a disease to appear, they could help us predict disease emergence and epidemics so that we can make informed decisions on early surveillance targets and intervention actions. Therefore, risk maps can support the planning, preparedness, and response to an epidemic or pandemic infectious disease or to an existing condition (chronic disease or endemic infectious disease). Risk maps can also provide information for surveillance programs, outbreak investigations, emergency management, and prevention and control programs, thus supporting resiliency solutions to health threats.

EO has proven its ability to detect risks of disease in the environment and the characterization and location of vulnerable populations (Section 2.5). Vulnerability indicators and environmental hazard maps are combined to produce risk maps so that public health decisions and actions can be targeted to those who need it and in the most efficient way.

The implementation of risk maps for risk monitoring on a continuous basis has been demonstrated in this book for applications on air quality and water quality (Section 2.3 and 2.4). These products require significant infrastructure and partnerships for big data management and process automation. The products generated meet important needs for public health – access to timely, accurate, and authoritative data and IT infrastructure to support effective evidence-based decision making.

To generate products effectively, public health organizations need to collaborate with space agencies and other organizations that provide access to EO missions and data streams. Integrating a variety of data sources (e.g. climate, socio-economic, environment) with these data streams for modeling and health systems to generate reliable and consistent results remains essential. Open data and data-sharing policies[1] and promotion of participatory approaches to generate and access geospatial information are important prerequisites, as is the collection of health-related *in situ* data to produce validated spatial analyses.

Spatial and analytical innovative solutions using EO in the domains of epidemiological analyses, risk modeling, surveillance, outbreak investigation, and emergency management have the potential to influence public health reactions and shift the epidemic curve toward prediction and prevention of diseases. EO-based information could support many time scales, such as daily and seasonal monitoring for early warning of diseases by forecasting models.

An increased capability of public health to conduct mapping and spatial analysis rapidly for decision making depends on the existence of the right combination of open-access data, methods, open-source software or code, tools, technologies, and infrastructure. Solutions for big data storage and analytics from EO and the generation of analysis-ready data to accelerate the computation of risk models could be facilitated by science and technology innovations such as artificial intelligence, machine learning algorithms, and data cubes. Infrastructures such as high-performance computing systems or cloud environments offer promising solutions that need to be explored. Although some satellites provide products that are ready for analysis and some EO satellite prototypes are designed to provide health data, the collaboration of users and science teams working on EO satellite missions could influence the development of innovative instruments that speak to public health needs.

Building skilled human resources is essential for effectively using EO data and deriving geospatial information from them. Expertise in EO image analysis and geomatics is vital for the development of risk maps. Methods to develop these skills are recommended for both communities to better collaborate, share knowledge and best practices, and support training and professional development in academia and government organizations. Finally, the development of a strong Community of Practice (CoP) with EO (remote sensing experts) and health (epidemiologists, modelers) and relevant sectors (entomology, biology, climatology, environmental sciences) is at the foundation of innovative actions. We have recommended several approaches to identifying and addressing global and national needs regarding health issues, including through interdisciplinary cooperation and joint activities, formal cooperative agreements and dialogue between sectors, and participation in current CoP such as the GEOHealth CoP.

In using this One Health framework together with EO data to examine real-time health challenges around the world, to support sustainable development, and to build human resiliency, we can make important contributions toward understanding and ultimately improving the health of not only humans but of animals and environmental systems alike.

Note

[1] http://www.earthobservations.org/open_eo_data.php (accessed 6 January 2022).

Appendix A Summary of Expert Presentations and Consultations

The following presentations delivered at the One Earth – One Health Workshop in Montreal, 2017[1] highlight the potential and value of Earth Observation (EO) technology for surveillance, prevention, control, prediction and/or forecasting, and public outreach activities. What became clear from these presentations was that there was a shared desire to help the United Nations (UN) meet their third Sustainable Development Goal – Good Health and Well-being (SDG 3), which endeavors to ensure healthy lives and promote well-being for all ages. The presenters focused on opportunities and challenges for development of applications and use of EO data in public health.

Strengthening National Capacities for Utilizing Satellite-based Earth Observation Data to Advance National Health-related SDG 3 Targets: A Conceptual Framework

Ramesha S. Krishnamurthy
Health Systems and Innovation Cluster, World Health Organization (WHO), Geneva, Switzerland; United Nations Committee on the Peaceful Uses of Outer Space (UN-COPUOS) Expert Group on Space and Global Health

Prevention, preparedness, response, and recovery require data sets. Data for evidence-informed decision making come from a variety of sources such as census, civil registration, vital event registry, surveys, health events and risks, individual records, health service records, resource tracking, and also satellite imagery, geospatial data, and base maps. These data sets for analysis use rapidly collected, extremely large volumes of both structured and unstructured electronic data through multiple data sources to answer complex questions that ordinarily cannot be answered using single data sets. There is a need to simplify big data for science and data for decision making. These data sets require a set of complex subsystems for the completion of a coherent health information system.

To address this issue, WHO has suggested a national framework for optimizing the combined use of Earth Observation satellite (EOS) data, routine health information data, and other remote sensing (RS) data to advance specific targets in SDG 3 at the national and subnational levels. The framework will address the mapping of country-specific health-related activities with EOS for the following: health technical areas, resource needs, and research agenda. The framework will also address the collaborative opportunities for space agencies and public health institutions. For countries to utilize EOS and RS data and develop national ownership of the framework, inter-sectoral collaboration, technical infrastructure, a competent workforce, and adequate finances are essential.

Group on Earth Observations (GEO) Activities and Health

Steven Ramage
GEO Secretariat

The GEO Secretariat is an intergovernmental partnership working to improve the availability, access, and use of EO for the benefit of society. GEO works to actively improve and coordinate global EO systems and promote broad, open data sharing. Membership in GEO is open to all Member States of the UN and to the European Commission. GEO has 113 Member Countries and 140 Participating Organizations (in 2022) that work to develop and implement projects and initiatives that solve global problems.

In the context of One Health and SDGs, GEO wants to connect public health and EO communities through the initiatives and flagships that comprise the GEO Work Programme, which includes more than 50 activities planned for 2020–2022. In particular, GEO would like to encourage inter-sectoral dialogue and collaboration via their Earth Observations for Health (EO-4HEALTH) initiative and other relevant activities to ensure the availability, open access, and reusability of EO data for public health purposes.

The intent is also to understand the interactions between different SDGs, expand the focus on health, and increase policy opportunities for geospatial data. To examine SDG interactions, GEO uses the International Council for Science report[2] as a guide for targeting interactions, from science to implementation. From the review of SDG interactions and other analyses, GEO wants to apply resilience systems analysis with a health impact lens.[3] A resilience systems analysis would provide key actors in the field with a shared view of the risk landscape that people face. This includes an understanding of the broader system that people need for their all-around well-being and an analysis of how the risk landscape affects the key components of this system.

The analysis would identify which components are resilient, which are not, and provide a shared understanding of power dynamics and how the use or misuse of power helps or hinders people's access to the assets they need to cope with shock and setback. Based on all of this, a shared vision of what needs to be done to boost resilience in the system and how to integrate these aspects into policies, strategies, and development efforts at every layer of society is needed. According to GEO, it is critical to understand the cause and effect of stresses and shocks in the past if we are to properly understand and prioritize future risks. The risk landscape brings both risks and opportunities: both should be considered when looking at how shocks and stresses affect systems and people.

To support SDGs, GEO put together the EO4SDG initiative, which helps identify which targets EO can help advance and what indicators are in development. This GEO initiative is in service of the UN 2030 Agenda for Sustainable Development.[4] It works closely with UN organizations such as the UN Committee of Experts on Global Geospatial Information Management (UN-GGIM) and is a partner in the Global Partnership for Sustainable Development Data (GPSDD) and the UN Sustainable Development Solutions Network's (SDSN) Thematic Research

Network on Data and Statistics (TReNDS). Indicators developed through EO4SDG's initiative related to health are for the moment focused on the quality of water-related ecosystems, land degradation, and climate change.

Along with the 2030 Agenda, GEO also supports the development of policy and the implementation of priorities related to the Paris Agreement for Climate Change and the Sendai Framework for Disaster Risk Reduction. Emerging work is also underway with respect to urban resilience and the New Urban Agenda.

The Global Earth Observation System of Systems (GEOSS) Platform[5] is a discovery tool with more than 400 million open EO data and information resources. Work is now in progress on a GEO Knowledge Hub, which is a framework for evolving the global work of GEO into a digital repository. The goal is to improve delivery of knowledge, services, and products. This framework proposes ways in which advances in hardware technologies, software tools, and cloud computing resources needed for handling, processing, and delivering big data from EO systems may be maximized to support evidence-based decision making. This hub is being created according to the GEO principles of supporting open data and open science.

CNES Activities in Tele-epidemiology: How Can Earth Observation Satellite Data Contribute?

Cécile Vignolles

Program Manager, Direction de l'Innovation, des Applications et de la Science/ Terre-Environnement-Climat, Centre National d'Etudes Spatiales, France (CNES); UN-COPUOS Expert Group on Space and Global Health

Tele-epidemiology consists of using space technology to study human and animal diseases that are closely linked to climate and the environment (i.e. transmitted by water, air, or vectors).

To meet the needs expressed by health authorities, the French Space Agency (CNES), in collaboration with its partners, has developed a deterministic/statistical approach to climate–environment–health relationships, producing EO products that are in line with the various spatio-temporal scales of the factors involved in the emergence and spread of infectious diseases. This tele-epidemiology conceptual approach is based on the study of key mechanisms favoring emergence and propagation of infectious diseases through a multi-disciplinary lens that includes the fields of environmental science, climate, social sciences, microbiology, entomology, and veterinary science. Tele-epidemiology includes experimental design mainly associated with field studies in order to obtain *in situ* data sets and to provide diagnostics such as the main physical and biological mechanisms at stake. It also includes RS products to monitor the environment and link epidemics with confounding factors. The last component of tele-epidemiology is dedicated to modeling for risk mapping, which involves building predictive models by combining *in situ* data and RS products derived from EO satellites, geographic data, and meteorological data to produce dynamic high spatio-temporal resolution risk maps. This conceptual approach has been applied successfully to different infectious diseases such as Rift Valley fever (Senegal), malaria (Dakar, Burkina Faso, French Guyana, Madagascar), dengue (metropolitan France, Martinique, French Guyana), and meningitis (Sahel region). CNES provides additional tools/services to public health actors to help them with disease surveillance and in the implementation of strategies for disease control.

Getting Ahead of the Curve: Using Earth Observations to Predict Health Risks

Juli M. Trtanj
One Health Lead, Climate and Weather Extremes Integration Lead, National Oceanic and Atmospheric Administration (NOAA) Climate Program Office

The NOAA One Health Group advances NOAA's science and services to inform health decisions through improved understanding of the links between environmental conditions and health outcomes and the delivery of useful prediction products, data, and tools. Thematic areas include: wildlife and zoonotic disease, air quality, heat (thermal extremes), vector-borne disease, water-borne illness, natural products, safe food, and the Arctic.

On the meteorological and climatological front, the National Integrated Heat Health Information System (NIHHIS) integrates existing forecasting and prediction capabilities for extreme heat at all time scales. The goal is seamless information that can inform decisions in many sectors at many time scales, such as early warning forecasts for emergency management that targets areas most vulnerable, and engaging in preparatory communications and planning for hospital administrators.

An example of NOAA's efforts regarding drinking water, recreational waters, and food contamination involves using new RS technologies to detect and predict high *Vibrio* spp. densities in oysters and coastal waters that could pose a risk to people. This demonstrates the use of EO to help create a real-time monitoring tool that involves public health, legislators, consumers, fishers, seafood processors, and aquariums. NOAA also provides tools for tracking and early warning systems for *Vibrio* from gene data to satellite and *in situ* data and is working on a cholera early warning system.

The ultimate goal of NOAA One Health is to contribute to changing the culture to prediction and prevention. The lead time provided by climate and ocean based predictive tools and information, observations, and monitoring gives us the opportunity to change a current epidemic curve to shift the whole scenario forward. This allows us to anticipate or be better prepared for the first case and to detect and monitor. By shifting the process forward in time, the opportunity to control an epidemic starts much earlier, allowing public health measures to reach a much greater number of people in a more timely, efficient way.

We benefit from integrated information systems for health, including the GEO Health Community of Practice, the Global Heat Health Information Network (GHHIN), and the establishment of regular dialogue between researchers, operations, and observers. These integrated information systems fix the gap in information and can help find solutions to public health issues, with early warnings that can lead to early actions.

Earth Observations for Health and Air Quality

John A. Haynes
Program Manager, Health and Air Quality, Applied Sciences Program, Earth Science Division, Science Mission Directorate, NASA, Washington, DC, USA

The NASA Earth Science Division and Applied Sciences Program has the mission to discover and demonstrate the innovative and practical uses of EO in the policy, business, and management decisions of an organization. The mission is expressed through: (i) applications to prove out, develop, and transition ideas into sustained applications of EO products in decision making; (ii) capacity building to build skills and capabilities in the USA and in developing countries so more countries can access and benefit from EO; (iii) mission planning to identify applications early in a mission's life cycle and integrate end user needs in mission design and development. The results of NASA Earth science research are in the areas of technology, missions and observations, data and archives, models, and predictions. These results feed decision management, policy decisions, forecasting, response, and recovery. The application areas are in various domains but include health and air quality. The objectives of health and air quality are to support the use of EO in air quality management and public health, particularly regarding infectious disease and environmental health issues. The area addresses issues of toxic and pathogenic exposure and health-related hazards and their effects for risk characterization and mitigation. The area promotes uses of EO data and models regarding implementation of air quality standards, policy, and regulations for economic and human welfare. The area also addresses the effects of climate change on public health and air quality to support managers and policy makers in their planning and preparations.

The health and air quality program supports a variety of different projects such as an early warning system for malaria risk in the Amazon and a prototype model for improved forecasts of respiratory illness hazard from red tide in the Gulf of Mexico. A formal health and air quality applied science team (HAQAST) developed studies on the impact of cook stoves on health and air quality using NASA satellites. They also studied the complex regional trends (2005–2015) in nitrogen dioxide NO_2 and sulfur dioxide SO_2 pollution from Aura's ozone monitoring instrument (OMI).

NASA also offers training through an applied remote sensing training program (ARSET) that provides end users with professional technical workshops and builds long-term partnerships with communities and institutions in the public and private sectors.

NASA's Earth Venture Instrument-3, a new multi-angle imager for aerosols, is targeted for launch in 2022. The mission objectives are to assess links between different air-borne particulate matter (PM) types and adverse birth outcomes, cardiovascular and respiratory disease, and premature deaths. The instrument is a multi-angle spectro-polarimetric imaging instrument for operation in a sun-synchronous Earth orbit to measure the particle types, sizes, concentrations, and geolocation of atmospheric aerosols.

Healthy Societies and Healthy Ecosystems: An Integrated Monitoring Approach for Biodiversity and Human Health

Michael Gill
Group on Earth Observations Biodiversity Observation Network (GEO BON) with the collaboration of EcoHealth Alliance

GEO BON's mission is to improve the acquisition, coordination, and delivery of biodiversity observations and related services to users, including decision makers and the scientific community. GEO BON's vision is to be a global biodiversity observation network that contributes to effective management policies for the world's biodiversity and ecosystem services. Their roles are to: (i) advance biodiversity modeling and prediction capabilities (species population and distribution; ecosystem extent, structure, function, and degradation; community composition; genetic composition; species traits); (ii) design and develop an interoperable biodiversity observation system at national, regional, and global scales and across terrestrial, marine, and aquatic systems; (iii) deploy and promote harmonized, state-of-the-art biodiversity observation tools and organizational partnerships with both space agencies and national, regional, and local *in situ* biodiversity observation networks.

To address biodiversity, GEO BON recognized a need for collaboration to: integrate health and environment data (different metrics and reporting systems); understand how biodiversity (including virus- and inter-species interactions) is affected by land use change for the purposes of management and prediction; and better manage and conserve ecosystems to improve public health. The goal is to change the approach from reactive to proactive with a focus on prediction, early detection, and prevention targeting of areas at greatest risk (e.g. areas of rapid change and/or wildlife/domestic animal chains).

Through its modeling working group, GEO BON brings together existing and new biodiversity observations to produce modeling tools that can infer and predict biodiversity change across time and space.

An example of collaboration between GEO BON and the EcoHealth Alliance, where health can be leveraged to address the underlying drivers of environmental degradation, can be found on its website.[6]

GEO BON and the EcoHealth Alliance are partnering to improve: (i) integrated *in situ* and remotely sensed monitoring systems that bring together biodiversity, ecosystem condition, and wildlife health observations; (ii) understanding of the link between changing ecosystems, socio-economic trajectories, and potential emergence of diseases; (iii) ability to predict (with improved and integrated models) potential emerging disease hotspots and to develop early warning systems; and (iv) policy guidance that results in positive outcomes for biodiversity, ecosystem services, and human health.

Pilot collaborations include: co-monitoring (sub)tropical areas experiencing rapid land use change (Brazil and China) using both *in situ* and RS data; integrating and refining test models for species distribution of known disease carriers with sociological data and wildlife origin/vector-borne infectious disease (e.g. making better predictive models for areas/conditions of future outbreaks);

upscaling and deploying models and monitoring systems to priority regions (e.g. hotspots) using GEO BON/EcoHealth Alliance networks, Future Earth's oneHEALTH project, and BON in a Box to reflect the joint program between the Convention on Biological Diversity and WHO; and integrating these methods in GEO BON national Biodiversity Observation Network frameworks and manuals.

One Health – Contribution of EO to Public Health Issues

Nicholas H. Ogden
Public Health Risk Sciences Division, National Microbiology Laboratory, Public Health Agency of Canada

What does emerging and re-emerging diseases mean? The former are diseases that emerge by short- and long-distance geographic spread and by "adaptive" emergence, i.e. the evolution of a microbe from one that is non-pathogenic for humans to one that is. Re-emergence is a state of increased risk from a pathogen due to environmental or human population changes that increase rates of contact between humans and pathogens. Most emerging infectious diseases (EIDs) are, or were originally, zoonoses, i.e. diseases of animals that are transmitted to humans.

The drivers of disease emergence were described – including changes to climate, habitat, and land use, biodiversity, and dispersion of pathogens locally and globally. Examples of the use of EO data in these fields were shown.

The main public health responses to EIDs are: model-based assessments of current and future risk; surveillance; and outbreak response. Examples of these, and how EO data are used to support these activities, were shown. These examples included: the use of EO data as climate, habitat, and socio-economic status proxies in predicting risk from emerging Lyme disease in Canada; the use in decision making for field surveillance campaigns; and the use in support of outbreak responses in difficult/remote regions such as responses to Ebola in Africa.

Future opportunities for applying EO data were described, including: use for enhanced assessments of risk; use of EO data to assist and eventually replace field surveillance; the use of high-spatio-temporal-resolution EO data in early warning/disease forecasting systems; and the integration of the "big data" fields of EO with modern genomics so that risk assessments target not just pathogen species but genotypes or particular genes.

EO data are being used in day-to-day public health activities; however, their full potential requires national and international collaborations in research, development, and implementation in public health.

Appendix B Spatial, Spectral, and Temporal Resolutions

The following two tables refer to the different classes commonly used to categorize the *resolution* of Earth Observation (EO) systems (Table AB.1) and their spatial, spectral, and temporal dimensions (Table AB.2).

Table AB.1. Classes of resolutions. (From: Kotchi *et al.*, 2015.)

	Resolution dimensions		
Class of resolution	Spatial (m)	Spectral	Temporal (days)
Very high	VSR: pixel ≤5	VER: NoB >30	VTR: RT ≤1
High	HSR: 5 < pixel ≤ 30	HER: 5 < NoB ≤ 30	HTR: 1 < RT ≤ 3
Medium	MSR: 30 < pixel ≤ 500	MER: 3 < NoB ≤ 30	MTR: 3 < RT ≤ 16
Low	LSR: pixel >500	LER: NoB ≤5	LTR: RT >16

VSR, very high spatial resolution; HSR, high spatial resolution; MSR, medium spatial resolution; LSR, low spatial resolution; VER, very high spectral resolution; HER, high spectral resolution; MER, medium spectral resolution; LER, low spectral resolution; VTR, very high temporal resolution; HTR, high temporal resolution; MTR, medium temporal resolution; LTR, low temporal resolution; NoB, number of bands; RT, revisit time.

Table AB.2. Examples of EO satellite sensor systems and their spatial, spectral, and temporal resolutions. (From: Kotchi *et al.*, 2015, with updates [http://database.eohandbook.com and https://space.oscar.wmo.int/spacecapabilities accessed 6 January 2022].)

Satellite	Sensor	Optical or SAR	Spatial (m)	Spectral (NoB)	Spectral range (nm)	Temporal (TR in days)
Geoeye-1	GIS-1	Optical	Pan 0.41 MS 1.64	5 in VIS/NIR	450–920	NA[a]
Worldview-2	WV110	Optical	Pan 0.46 MS 1.84	9 in VIS/NIR	400–1040	NA[b]
SPOT-5	HRG	Optical	Pan 2.5 MS 10 SWIR 20	5 in VIS/NIR/ SWIR	490–1750	26/3[c]
SPOT-6/7	NAOMI	Optical	Pan 2 MS 8	5 in VIS/NIR	450–890	30/3 days[d]
Landsat-5	TM	Optical	MS 30 TIR 120	7	450–2350	16

Continued

Table AB.2. Continued.

Satellite	Sensor	Optical or SAR	Spatial (m)	Spectral (NoB)	Spectral range (nm)	Temporal (TR in days)
Landsat-7	ETM+	Optical	MS 15 TIR 60	8		16
Landsat-8	OLI	Optical	MS 15 TIR 100	11		16
Terra-Aqua	MODIS	Optical	250 to 1000	36	459–2155	16
Suomi NPP	VIIRS	Optical	750	22	VIS, NIR, SWIR, MWIR, LWIR	1
Sentinel-5P	TROPOMI	Optical	7 km	4	UV, VIS, NIR, SWIR	1
NOAA	AVHRR	Optical	1100	6	580–12,500	NA
Sentinel-2A and 2B	MSI	Optical	10 to 60	13 in VIS/NIR/SWIR	443–2190	5 to 10[e]
Pléiades	HiRI	Optical	PAN 0.7 MS 2.8	5 in VIS/NIR	450–900	26/2[f]
Envisat	ASAR	SAR	300	1	C-band	35
ALOS	PALSAR	SAR	6.25 to 100	1	L-band	14
RADARSAT-2	SAR	SAR	1 to 100	1	C-band	24
TerraSAR-X	SAR	SAR	0.5 to 40	1	X-band	4 to 7
CosmoSkyMed	SAR	SAR	1 to 100	1	C-band	16
Sentinel-1A and -1B	SAR	SAR	5 to 20	1	C-band	12 to 5
RCM	SAR	SAR	1.3 to 5000	1	C-band	1

ALOS, Advanced Land Observing Satellite; ASAR, Advanced Synthetic Aperture Radar; AVHRR, Advanced very high-resolution radiometer; ETM+, Enhanced Thematic Mapper plus; GIS, geographic information system; HiRi, High Resolution Optical Imager; HRG, High Resolution Geometric; LWIR, long-wave infrared; MODIS, Moderate Resolution Imaging Spectroradiometer; MS, multispectral; MSI, multispectral instrument; NIR, near infrared; NPP, National Polar-orbiting Partnership; MWIR, mid-wave infrared; NA, not available; NAOMI, New AstroSat Optical Modular Instrument; NoB, number of spectral bands; OLI, Operational Land Imager; PAN, panchromatic; PALSAR, Phased Array L-band Synthetic Aperture Radar; RCM, RADARSAT Constellation Mission; SAR, synthetic aperture radar; SPOT, Système Pour l'Observation de la Terre; SWIR, short-wave infrared, TIR, thermal infrared; TM, Thematic Mapper; TR, temporal resolution; TROPOMI, Tropospheric Monitoring Instrument; UV, ultraviolet; VIIRS, Visible Infrared Imaging Radiometer Suite; VIS, visible.

[a]Global coverage in 6 months, in daylight. One area can be observed in as few as 4 days with strategic pointing.
[b]Global coverage in 6 months, in daylight. One area can be observed in as few as 3 days with strategic pointing.
[c]Global coverage in 26 days, in daylight. One area can be observed every 3 days with strategic pointing.
[d]Global coverage in 30 days. Minimum revisit time for a specific area is 3 days.
[e]Global coverage in 10 days, in daylight with 1 satellite (or in as few as 5 days with the 2 satellites).
[f]Global coverage in 26 days, in daylight. One area can be observed every 2 days with strategic pointing.

Notes

[1] https://crss-sct.ca/conferences/csrs2017/one-earth-one-health-workshop/ (accessed 6 January 2022).
[2] https://www.icsu.org/publications/a-guide-to-sdg-interactions-from-science-to-implementation (accessed 6 January 2022).
[3] https://www.oecd.org/dac/Resilience%20Systems%20Analysis%20FINAL.pdf (accessed 6 January 2022).
[4] https://www.earthobservations.org/documents/publications/201703_geo_eo_for_2030_agenda.pdf (accessed 6 January 2022).
[5] www.geoportal.org (accessed 6 January 2022).
[6] https://www.ecohealthalliance.org/ (accessed 6 January 2022).

References

Chan, M., Kazatchkine, M., Lob-Levyt, J., Obaid, T., Schweizer, J. *et al.* (2010) Meeting the demand for results and accountability: A call for action on health data from eight global health agencies. *PLoS Medicine* 7. DOI: https://doi.org/10.1371/journal.pmed.1000223.

Jones, K.E., Patel, N.G., Levy, M.A., Storeygard, A., Balk, D. *et al.* (2008) Global trends in emerging infectious diseases. *Nature* 451, 990–993. DOI: https://doi.org/10.1038/nature06536.

Kotchi, S.O., Brazeau, S., Turgeon, P., Pelcat, Y., Légaré, J. *et al.* (2015) Evaluation of Earth Observation Systems for estimating environmental determinants of microbial contamination in recreational waters. *IEEE Journal of Selected Topics in Applied Earth Observations and Remote Sensing* 8. DOI: https://doi.org/10.1109/jstars.2015.2426138.

Proença, V., Martin, L.J., Pereira, H.M., Fernandez, M., McRae, L., Belnap, J. *et al.* (2017) Global biodiversity monitoring: From data sources to essential biodiversity variables. *Biological Conservation* 213, 256—63. DOI: https://doi.org/10.1016/j.biocon.2016.07.014.

World Health Organization (WHO) (2004) Joint inter-ministerial policy dialogue on ehealth standardization and second WHO Forum on eHealth Standardization and Interoperability. Geneva Switzerland. Available at: http://www.who.int/ehealth/events/final_forum_report.pdf (accessed 20 June 2021).

Index

Note:-Page numbers in **bold** type refer to **figures** and Page numbers in *italic* type refer to *tables.*